MW01527348

SUTHERLAND QUARTERLY

SQ

Sutherland Quarterly is an exciting new series
of captivating essays on current affairs
by some of Canada's finest writers,
published individually as books
and also available by annual subscription.

The Rise and Rise of Canada's
Corporate Welfare Bums
At the Trough

Laurent Carbonneau

Five Days of Hell in a Rocky Mountain Paradise
Jasper on Fire

Matthew Scace

Canadians Versus Their Banks
Fleeced

Andrew Spence

Governing in Troubled Times
**Justin Trudeau
on the Ropes**

Paul Wells

SUBSCRIBE ONLINE AT
SUTHERLANDQUARTERLY.COM

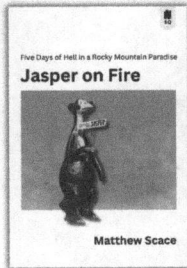

AT THE TROUGH

AT THE TROUGH

The Rise & Rise of Canada's
Corporate Welfare Bums

LAURENT CARBONNEAU

SQ

sh.
SUTHERLAND HOUSE

Sutherland House
416 Moore Ave., Suite 304
Toronto, ON M4G 1C9

Copyright © 2025 by Laurent Carbonneau

All rights reserved, including the right to reproduce this book or portions thereof in any form whatsoever. For information on rights and permissions or to request a special discount for bulk purchases, please contact Sutherland House at sutherlandhousebooks@gmail.com Sutherland House and logo are registered trademarks of The Sutherland House Inc.

First edition, March 2025

We acknowledge the support of the Government of Canada.

Manufactured in Canada
Cover designed by Luisa Galstyan, Shalomi Ranasinghe, and Jordan Lunn
Book composed by Karl Hunt

Library and Archives Canada Cataloguing in Publication
Title: At the trough : the rise and rise of Canada's corporate welfare bums / Laurent Carbonneau.
Names: Carbonneau, Laurent, author.
Identifiers: Canadiana (print) 20250117347 | Canadiana (ebook) 20250117371 |
ISBN 9781998365579 (softcover) | ISBN 9781998365586 (EPUB)
Subjects: LCSH: Subsidies—Canada. | LCSH: Industrial policy—Canada. |
LCSH: Canada—Economic policy. | LCSH: Canada—Social conditions.
Classification: LCC HC120.S9 C37 2025 | DDC 338.971—dc23

ISBN 978-1-998365-57-9
eBook 978-1-998365-58-6

Sutherland Quarterly, **Issue 9**
Editor – Ken Whyte
Managing Editor – Shalomi Ranasinghe
Associate Editor – Leah Ciani
Marketing Director – Serina Mercier
Publicist – Sarah Miniaci
Subscription Price: $67.99 CAD (includes HST) | Single Copy Price: $17.95 USD / $19.95 CAD
For submissions and more information, e-mail us at submissions@sutherlandhousebooks.com.

CONTENTS

To my teachers

INTRODUCTION

In April 2023, a gaggle of ministers of the Crown descended on the small southwestern Ontario community of St. Thomas bearing wonderful news.

Canada's Liberal Prime Minister Justin Trudeau and Ontario's Progressive Conservative Premier Doug Ford were about to announce that Canadians would be awarding German auto giant Volkswagen as much as $13 billion in subsidies over the next decade to build a gigantic new electric car battery plant in the town of around 40,000 people.

The announcement must have caught the attention of the Dutch-headquartered automaker Stellantis, which produces Chryslers and Jeeps. It promptly renegotiated a previously announced subsidy of $1 billion from the Canadian and Ontario governments for its planned Windsor electric vehicle battery plant. It, too, would receive $15 billion, matching Volkswagen's terms.

The deals—and these really were deals, driven by an entrepreneurial federal minister's desire to compete with the United States and land Canada a starring role in the global electric car value chain—are the biggest commitments to private companies in Canadian history. But they were far from being our governments' first big, transformational bets to reshape the economy.

Everywhere one turns in Canada, one sees governments enthusiastically supporting businesses. Often, the gifts to lucky capitalists come with a personal delegation of beaming ministers bearing oversized novelty cheques. Even more often, support flows via eye-wateringly generous tax breaks for investments or activities like research and development that go unappreciated by the voting public.

Yet, as this public money flows faster than ever to businesses, Canadians are getting less and less value from their public services. Many programs overseen by the same governments writing those big cheques are rapidly deteriorating and approaching crisis points in their basic functionality.

This is not a new problem. Canada has eagerly doled out public money hand over fist to businesses since the moment four colonies became one country in 1867, for reasons ranging from outright corruption to well-intentioned incompetence. The graft is interesting in a sordid way, but it is the century and a half of well-intentioned, supposedly strategic largesse that is the genuinely riveting story of corporate welfare in this country.

Whenever ministers arrive somewhere with a cheque, they are always proud to claim that *this* will be the "investment" that sets Canada on the path to the top of the league tables in important global indicators of productivity and economic success. And somehow, the country stubbornly lags behind its peers in the global north. Italy is rightly regarded as a governance basket case, wracked by instability and sclerotic institutions. Yet, to take one indicator often seen as an important sign of economic dynamism, Italy recently passed Canada in rankings of business investment in research and development.

The question is obvious. How has Canada spent so much public money to achieve so little for so long?

There are a range of views for and against the widespread practice of governments supporting business. Some decry any and all spending of this nature as unearned spoils, corporate welfare of the worst kind, reflecting the corruption of the pristine free market by the demagogic carnival-barkers who invariably sit as our elected representatives. There is a left-wing mirror image of this view that maintains Canada is actually "three mining companies in a trench coat" who have captured our government and trained it to do their bidding.

Self-styled pragmatists maintain that government support for business is a competitive global sport that can be played well or badly, much like international soccer, never mind that it often looks more like the ugly bureaucratic and political machinations that surround international soccer. There is also the earnest view that while government support for *business* is usually misguided and misspent, the best tonic to these unearned windfalls for the rich is broader government involvement in *the economy*.

While I understand the frustration and respect both the crankiness and ideological purity of the first two groups, in the absence of the reasonable prospect of a socialist revolution tomorrow, I find my analysis ends up somewhere between the latter two camps.[1] Canada's reality is that as a small, open economy in a global capitalist system that prizes relatively free flows of investment across national borders, governments have to be smart in ensuring that capital flows end up benefiting the country, by raising productivity and building up wealth to allow for a generous and democratic social state. The long-run result of stagnation is to be a source of cheap labour and learning to live with the highly unequal political economy that comes with that.

The corporate welfare state, that policy apparatus of business subsidies Ottawa has built since Confederation, has not delivered on its stated objectives of promoting economic growth and social welfare. Nevertheless, it keeps growing and, in fact, we've recently passed an important threshold. Since 2019-2020, Canadians have been giving away more than 50 cents of every dollar collected in corporate income taxes right back to businesses.

It's important to understand where we are, how we ended up here, and where we can go in the future. Canadians are facing serious crises in critical public services like healthcare and childcare, and social goods like housing. Does it really make sense for us to give Canadian businesses back more than half of what they pay in taxes when so many people can't find a family doctor or afford a home?

This book will provide a whirlwind tour of the Canadian landscape of corporate welfare and business subsidies in all of its glory. We'll start with a snapshot of today's subsidizing state and go back to see how we got

---

1 Should the revolution come, however, let the record show that I always thought Canada was three companies in a trench coat.

here. Then, we'll return to the present armed with context, and suggest how we redirect corporate giveaways to make real differences to address both Canada's big economic challenges and the social challenges that Canadians live with every day instead.

CHAPTER ONE

TODAY'S SUBSIDIZING STATE

Let's take a quick look at the busy fall 2022 schedule of the federal industry minister, François-Philippe Champagne. His department, which has had many names but is now called Innovation, Science and Economic Development (ISED), has been a major player in the subsidies business since its creation.

In mid-October, Minister Champagne joined Prime Minister Trudeau in Sorel, Quebec, for a press conference with global mining giant Rio Tinto. They announced that the federal government would contribute $222 million to efforts to make the company's local titanium and scandium processing plant more environmentally friendly. This funding was pitched as a key initiative of the federal government's push to establish a homegrown electric car battery industry.

A few days later, Champagne, the prime minister, and Premier Ford hit a stage with senior officials of Finnish telecommunications giant Nokia in suburban Ottawa to announce that the federal, provincial, and municipal governments would contribute $72 million between them to the company's planned expansion in the national capital.

In early November, Minister Champagne was in Edmonton with his federal colleagues and counterparts from the provincial United Conservative government to unveil $300 million in federal support for a $1.6 billion investment by Air Products Canada in a hydrogen production facility in Alberta's capital.

In just under a month, the federal government had announced, in partnership with two often-hostile provincial conservative governments, just over half a billion dollars in funding for large projects by established companies, two of them arguably global household names.

It was a big fall for the innovation minister, big companies, and big cheques. But it was hardly the only largesse distributed by the federal government in 2022. Canada's small businesses had been promised $4 billion in grants and loans through ISED and the Business Development Bank of Canada (BDC) in March to help get their products and services online. This was the new Canada Digital Adoption Program.

Hundreds of subsidies, incentives, and supports for private businesses, large and small, foreign and domestic, were announced by the federal and provincial governments in 2022. And it was only a few months into 2023 when Trudeau's government announced the unprecedented package of handouts to automakers Volkswagen and Stellantis with which we began our story. Things have continued apace from there. The Strategic Innovation Fund, a support program for big, technology-intensive projects, rolled out another $7.5 billion in announced funding through the rest of 2023, and billions in clean manufacturing tax credits came online in spring of 2024.

If you ask staffers or ministers in the Trudeau government to name the one achievement in which they take the most pride, many will point to the Canada Child Benefit (CCB), created soon after they formed government in 2015. I don't blame them. The program has taken 300,000 children out of poverty.[2] The benefit was expected to cost $26 billion in 2022, a substantial chunk of change. Yet, in 2023, the cost of the CCB was outweighed by the mass of subsidies, tax incentives, and preferential access to capital on offer to corporations from the government. Would many proud Liberals put the renaissance of the corporate welfare state at the top of their list of accomplishments?

* * *

2 They have since decided it needed to be augmented with a universal affordable childcare system as many people argued was necessary when they were first elected. I will elect to express no bitterness and instead chalk it up to personal growth on their part.

The first thing to understand about the last decade or so of subsidies is that they have exploded in magnitude. To present this as accurately and completely as possible, I'm hewing close to the methodology used by a former Department of Finance official, John Lester, in a 2018 paper that totalled federal and provincial subsidies as they existed in 2014-15. Lester defines business subsidies as all support to business with an economic outcome in mind, excluding supports that have social welfare, promotion of culture, or other goals as their primary objective. (That's not to say that many of these excluded subsidies don't make their way to corporate coffers: film and television production got $360 million worth of federal tax credits in 2023, and the federal government transferred a further $190 million through the Canada Media Fund.)

Leaving out measures intended to reduce carbon (and other greenhouse gas) emissions was, I think, a defensible position in Canada in 2014-15. In the age of green industrial policy, I think it is a bit of a tougher sell. I've chosen nevertheless to keep them out of my math here for the sake of consistency; we'll discuss them later on. Similarly, we'll leave out pandemic support programs, which would completely dwarf and render irrelevant any pre-pandemic comparative data.

Totalling up program spending, implicit subsidies through Crown corporations, and tax credits and deductions, Lester found that in 2014-2015, the last full year of Stephen Harper's Conservative government, the federal government doled out $14 billion in subsidies to business. That works out to $390 for every person in Canada.

Starting in 2017-2018, we can observe rapid growth in these figures. From 2015 to 2018, the Liberals added $3.3 billion to federal business subsidies. They more than matched that over the course of the next year with $3.6 billion more in new money. In 2019-2020, subsidies grew again by $2.5 billion, followed by another increase of $2.5 billion in 2020-2021 and another of $1.8 billion in 2021-2022. More recent estimates are even higher. See table on the next page.[3]

---

3 This table combines my own analysis from public accounts data and Lester, 2024.

Federal business subsidies, excluding climate-related and non-economic measures

Year	Program Spending (percent share)	Tax Expenditures (share)	Implicit Subsidy through Government Business Enterprises[4] (share)	Total (billions)	Dollars Per Capita
2017-2018	$4.2 (24%)	$9.9 (57%)	$3.2 (18%)	$17.3	$467
2018-2019	$3.9 (18%)	$13.8 (66%)	$3.2 (15%)	$20.9	$556
2019-2020	$4.7 (20%)	$14.1 (60%)	$4.6 (19%)	$23.4	$615
2020-2021	$7.4 (28%)	$14.8 (57%)	$3.7 (14%)	$25.9	$677
2021-2022	$8.6 (31%)	$15.6 (56%)	$3.5 (12%)	$27.7	$701
2022-2023				$29	$734
2023-2024	$10.2 (31.1%)	$18 (54.7%)	$4.6 (14.1%)	$33	$833

The long story short is that in nine years, over the lifetime of the incumbent government, federal subsidies to business have more than doubled through the introduction of over 100 new programs. Every Canadian has gone from paying just over $310 to businesses large and small to over $800 per year in 2023-24.

As mentioned, this does *not* include two important families of subsidies: measures intended to reduce businesses' carbon emissions or those that were intended to support business through the pandemic. Adding numbers from climate-related business policies to our annual tally makes them bulge noticeably, up to $40 billion in 2023-2024, or $1,007 for each Canadian. And as more announced measures of all kinds come online, subsidies will reach $50 billion every year by 2027-2028. Assuming no more get announced between now and then, of course—a fairly unlikely scenario.

If climate-related programs make the numbers bulge, the COVID support programs take them right off the charts. The Canada Emergency

4   See methodology established in Lester, "Business Subsidies," 2015; assuming 8.25% social opportunity cost of capital (Lester's footnote reads 10% but he actually uses 8.25%).

Wage Subsidy (CEWS) alone doled out $100 billion from March 2020 to October 2021, or $20 billion more than the Canada Emergency Response Benefit (CERB). The other flagship business support through the pandemic, the Canada Emergency Business Account (CEBA), approved $49 billion in interest-free loans, of which about 75 percent was paid back as of the final deadline in January 2024. Of course, it's hard to begrudge the pandemic measures. The circumstances were unprecedented, these programs were one-offs, and avoiding a massive wave of bankruptcies and layoffs has so far kept Canada out of a bruising post-pandemic depression. No one thought in 2020 that 2023's biggest problem would be that the economy is *too* hot.

Even without the pandemic, however, the federal government has spent most of a decade making choices that have resulted in Canadians handing over more than 50 cents of every corporate income tax dollar right back to corporate Canada. You might expect that with all this support, Canada has roared to the top of those rankings of global indicators of productivity and economic success, or at least caught up to Italy. But no. Our economic performance is worse. We have warning lights popping up around all of our important indicators. Productivity, or the amount of output for every hour worked, has actually declined since 2018.

Worse, while we've been shovelling money at corporations, our social services have declined. There are two million more Canadians living without a family doctor than there were in 2019. Housing has become wildly unaffordable, with the Bank of Canada's Housing Affordability Index reaching heights unseen since the excruciating double-digit interest rates of the early 1990s. And that marquee Liberal social program, universal access to subsidized childcare spaces, is hitting bumps with staff shortages and long-term funding gaps that leave its future in doubt.

We'll return to look at the details of today's subsidizing state in more detail later on, and our options for tackling these big economic challenges. But to really understand how and why Canada has built such an incredible corporate welfare state, we have to look at the full story, from Confederation to the present. Corporate welfare is as much a part of the story of building Canada as fur trappers, the Golden Spike and the Avro Arrow.

CHAPTER TWO

RAILWAYS AND PROTECTION: CORPORATE WELFARE FROM MACDONALD TO LAURIER

Canada's history is taught to schoolchildren as the slow knitting together of hinterland colonies by settlers lured to a vast frontier by the lustre of fur, the sparkle of cold seas teeming with fish, and the glitter of gold— prizes deemed worthy of the genocidal displacement of the peoples who were here first. This is very unlike how I was taught American history in the United States, where I grew up. There, a procession of firm-jawed presidents descended from the heavens to guide the world's oldest and greatest republican experiment to its manifest destiny.

I note this because I find it interesting that Canada, a country that has a much more ambiguous relationship in both public and private life about the desirability of making money, does not indulge in American-style romanticism about how we carved this nation out of the wilderness with our bare hands. Our greatest historians are less nationalist mythmakers or sycophantic court historians, and more fussy bean counters.

This furtive, penny-pinching side of us has fuelled our national obsessions with problems of economic development. We are as desperate to keep up with the Americans as we are to keep ourselves out of their hands. We are always on the lookout for new paths to prosperity, and always willing to open the public purse to private businessmen who claim they have the cure for what ails us.

The North American colonies that became Canada spent the first half of the nineteenth century tightly tied to the global British imperial economy. Then came a new era of free trade and a chain of events that ended with the birth of Canada.

The UK Parliament repealed its Corn Laws in 1846. The high tariffs that had protected British grain farmers and exporters in the colonies from foreign competition disappeared. In 1851, the last of the preferential imperial tariffs that had given its colonies favoured access to British markets was dispatched. This led to economic turmoil in the colonies. Mobs in Montreal burned the colonial parliament building. There were calls to annex Canada to the United States.

Realizing that it had to act to prevent its colonies from falling into the arms of the US, Britain opted to compensate them for their loss of preferred access to imperial markets by negotiating a 1854 trade reciprocity treaty with the United States. This tied together a continental market, at least for primary goods like fish, grain, lumber, coal and meat, while maintaining British hegemony north of the American border.

A period of relatively free trade ensued, coinciding with rapid population growth in the colonies. Their population more than doubled between the 1851 and 1871 censuses, from 2.4 million to 5.4 million. Then the American Civil War ended and a more protectionist American administration opted to abrogate the reciprocity treaty.

Two serious trade shocks in a single generation were difficult for the political class in the colonies to weather. Its members realized that if British North America was to have a future as an entity separate from the United States, they needed to promote internal colonization, especially in the form of westward expansion, to grow a more robust domestic market and stave off annexationist pressure from the south. These external pressures and a desire for better economic integration among the British North American colonies contributed to the decision made by Quebec, Ontario, New Brunswick, and Nova Scotia to join together in Confederation. The new Conservative federal government of Sir John A. Macdonald now faced the questions of how to integrate disparate provincial and class interests and, after British Columbia's accession to Confederation in 1871, how to forge a continental railway link to its Pacific province within a promised ten years.

These were difficult problems. To address them, Macdonald's government used public funds to heavily subsidize the private business interests that would build the railroad. These massive subsidies went hand-in-hand with high tariffs to spur domestic industry, the so-called National Policy. Thus began the proud Canadian tradition of dependent development that Canada is still trying to correct today. Spurred by a mixture of jealousy and fear of American growth and expansion, the tradition developed in a sporadic un-strategic manner, leaving ordinary working people worse off than if the government had done nothing at all.

* * *

What are we talking about when we talk about railways? Not the railways of today, which play largely marginal and invisible roles in our lives and economies. Nineteenth-century railways were simultaneously the premier transportation, telecommunications, and industrial development corporations of their day. Companies like the Canadian Pacific Railway were enormously significant political players whose executives dealt directly with prime ministers and major European and American financiers as peers (and occasionally as superiors).

The railroads were accountable primarily to distant, mostly British shareholders who were looking for steady and growing profits. They had fingers in every pie across the economy and were unconstrained in any meaningful way by regulation or law. Communities and industries lived and died by their decisions and their freight rates, their control of infrastructure like grain elevators and ports, and their incredible political influence. The easiest way to conceive of the significance of the big railways to the pre-automobile economy is to imagine if a national telecom, a bank, a national airline, and major auto manufacturer were all one company today; it is not hard, by extension, to imagine how our political system would go out of its way to accommodate such a class of giants.

By the late 1860s, Canada's premier railroad company, the Grand Trunk, was nearly insolvent and a shareholder revolt led to a refocus on dividends rather than undertaking grand nation-building projects. This left Macdonald's government without a partner to build out its promised link to the Pacific. Corporate intrigue ensued, culminating in Sir Hugh Allan's Canadian Pacific company being granted a federal charter in 1873 to build our first transcontinental railway, backed by a parliamentary gift of 50 million acres of land and a $30 million loan to be paid back in instalments.

The investment pitch for Allan's railway, which chose a Canadian route to the west coast rather than a more efficient international route through the United States, was that considerable settlement would quickly follow along the line. The project was probably doomed to failure as envisioned, planned, and subsidized, but Allan was spared the ignominy of failure. Along the way to securing his charter, he and his henchmen had illegally poured funds into the Conservative Party's coffers[5] and thrown American backers overboard to allay the government's concerns over foreign control of the line. The spurned partners paid Allan back by publicizing the bribes, leading to the fall of the Conservative government in the Pacific Scandal mere months after the CPR was granted its charter and before any subsidies had been distributed or construction begun. The year 1873 also saw a financial panic and the beginning of a massive depression that crippled the global economy for years.

Alexander Mackenzie's Liberal government came to power in 1873. It was not closely tied to either the Grand Trunk or Canadian Pacific factions of Canada's nascent rail-baronial class. He slow-walked railway development in light of the 1873 crash and subsequent slow recovery, opting to build bits and pieces as public works or via modest subsidized contracts. In light of the depression and dim prospects for quick scores, little progress was made.

5 There is a small but important to make here about the ethics of public office in the late 19th century: what we would call today conflicts of interest were not perceived as problems - scores of MPs and even ministers had active business interests, including in railways, that they were not shy about promoting - but outright bribery was still frowned upon.

When the ongoing depression brought Macdonald back to power in 1878, his government quickly moved to find a new partner on a transcontinental railway and introduce the National Policy of protective tariffs to promote domestic industries.

In 1880, a new rail syndicate headed by the expatriate Scot, George Stephen, won a new contract from the government to build the Canadian Pacific Railway. Like the abortive Hugh Allan project, it was generously subsidized: the company was given 25 million acres of land (a smaller package of choicer lands than the 50 million set aside for Allan), $25 million in cash, tax breaks for construction and operation, 1,100 kilometres of government-owned track (worth approximately $38 million). The deal also extinguished aboriginal title along the route (valued at $30 million). Importantly, the terms of the deal prohibited for decades any competitors from emerging to the line's south without the CPR's approval. The scale of the giveaway was enormous. It is difficult to put a modern dollar figure on it, but by way of comparison, federal expenditures in 1881 otherwise totalled only around $33 million. Provincial and local governments also chipped in.

In the end, the CPR's construction costs added up to $150 million, $88 million of which came from government. The public funding itself came from bond sales, financed, in turn, mostly by the new protective tariffs, which were regressive taxes on virtually every Canadian. The private funding involved sophisticated financial engineering. Rather than seeking bond financing at fixed rates of interest—the sort of obligations that had bedevilled other railways—the line's owners instead issued equity stakes that came with generous dividend terms, leaving the railway with an effective interest rate of 12.5 percent. This was actually much higher than they would have had to pay bondholders, but it insulated the owners from loss of control in the event of being unable to cover interest charges.

The structure of the railway's public and private financing formed an extractive system with three salient features. First, the CPR's owners were protected from potential loss of control to private bondholders through their use of dividend-bearing equity stakes, leaving the bond-raising to the public sector. Second, Canadian taxpayers were on the hook for the cost of the government bonds issued to British and American financiers, and paid it back on the railway's behalf through higher prices on all kinds of useful, everyday goods thanks to the new National Policy tariffs. Third, the owners were paying themselves generously with dividends and using their public funding to buy up feeder lines in an effort to kill their main competitors in central and eastern Canada to complement their chartered monopoly in western Canada.

Twice the CPR had to go back to the government for loans. Its finances were stretched by inflated construction costs, largesse to its owners, and the cost of their efforts to buy out their competition. The first time, in 1884, George Stephen announced to all who would listen in Ottawa that the sky would fall if the CPR did not receive more federal money. Part of the duly granted loan went to pump up the equity of a land speculation company, also owned by the CPR's directors. That company ultimately failed, but the owners got another break when the government agreed to buy back land grants at prices far above the depressed market value.

The second CPR bailout loan was eased by the railway's role in ferrying troops west to Saskatchewan to put down Louis Riel's 1885 rebellion—a rebellion in which several First Nations joined Riel's Métis rebels in part because the federal government, in the face of declining tariff revenues amid a short recession, had opted to "cut back allocations to minor priorities like the fulfillment of its treaty obligations in favour of maintaining the flow of public money to the CPR."

Further subsidies followed for small, private complementary and feeder lines north of the main CPR route in the west and all over the east: the government doled out $3,200 for each mile of track. These smaller lines were largely unviable, poorly constructed, or otherwise useless, and most wound up in possession of the CPR, with the original owners— often Macdonald's political allies—well-compensated, both directly and indirectly.

The last, golden spike in the CPR was hammered at Craigellachie, BC in November 1885. You've probably seen the picture. Was the CPR worth building at such absurd public expense? Even leaving aside the incredible quantity of money involved, the quality of the spending was abysmal. The CPR's financial engineering focused on maximum returns to shareholders rather than efficient construction and operation of the transcontinental line. The hoped-for wave of agricultural settlement did not follow immediately, putting a huge dent in the value of the premium the government paid to secure construction as quickly as feasible. The big winners were the CPR's investors and the owners of Canadian federal debt, which is to say, American and European financial institutions. It was an early object lesson in the noxious nexus of political excitability, skillful lobbying, and grand national visions. Unfortunately, it was a lesson left unlearned.

Karl Marx, a perceptive observer of human foibles, wrote, following Hegel, that history repeats itself, first as tragedy, next as farce. If the CPR was a blood-soaked tragedy[6] that came decades too early to be truly worth the price paid, Wilfrid Laurier's government's efforts to build out the Canadian Northern and Grand Trunk transcontinentals were the farce.

In 1899, Sir Wilfrid Laurier's Liberals soberly chose to end the practice of blanket subsidies for new railroads. The established players greeted this with delight. Their upstart competitors largely would have to fend for themselves. But a few years later, amid a boom driven by western agriculture, the Laurier government decided in a fit of national pride and optimism that what Canada needed was not one but two new transcontinental lines.

Laurier found dance partners in the venerable Grand Trunk and the Canadian Northern railways, which were both competing to build transcontinental alternatives to the CPR. An attempt by the government to bring the two to merger fell apart in 1903, leaving the government with little political choice but to support both with subsidies and land grants. The scale of Laurier's largesse made Macdonald "look like a miser." Between Macdonald, Mackenzie, and the series of minor Conservatives that preceded Laurier, the federal government had spent a little over

---

6 I have not touched on it in the text above, but hundreds of workers died building the railroad, and of course the role of the railways in dispossessing Indigenous people of their land has been well told in other places. James Datschuk's *Clearing the Plains* is an excellent recent treatment.

$200 million on railway subsidies. Laurier and his successor, Sir Robert Borden, spent nearly $400 million. The construction of the two competing northern railways was even more farcically corrupt than the CPR's had been, and their owners had even less skin in the game. They saddled the new lines with enormous and expensive debt that drove them to seek quick profits from extractive industries along their routes.

By the eve of the Great War, both the Canadian Northern and the Grand Trunk, despite all their public support, were tottering on mountains of debt from a decade of frenzied. Canada found itself in the 1920s with 16,000 kilometres more track than it actually needed. The federal Railway Board's refusal to allow Canadian Northern to raise rates during the war helped bring its finances to a crisis. The government realized it had an issue of too-big-to-fail financial exposure on its hands. In the end, the Borden government bailed out both of the bloated, debt-ridden, overbuilt northern lines, and in 1923 consolidated and nationalized and them as Canadian National Railways.[7]

* * *

The railways, serving as arteries of communications and settlement, were one important leg of Canada's efforts to grow its population, secure a robust home market, and ensure its own continued existence distinct from the United States. The other was leg was the long-standing National Policy, which subsidized domestic manufacturers by imposing tariffs on the import of foreign goods.

Canada had tried in vain to re-negotiate American trade agreements in 1871 under Macdonald) and in 1874 under Mackenzie. Hundreds of thousands of Canadians were leaving for the United States and its booming industrial economy. Macdonald spoke powerfully on the campaign trail in 1878, amid the lingering downturn from the global Panic of 1873: "We have no manufactures here. We have no work-people;

7 A move that estranged the Tories from the railway financiers for a generation.

our work-people have gone off to the United States. They are to be found employed in the Western States, in Pittsburgh, and, in fact, in every place where manufactures are going on. . . . If we had a protective system in this country, if we have a developed capital, we could, by giving our manufacturers a reasonable hold on our home trade, attain a higher position among the nations."

Upon his return to government after that campaign, Macdonald's first budget bill declared that "the welfare of Canada requires the adoption of a National Policy, which, by judicious readjustment of the Tariff, will benefit and foster the agricultural, the mining, the manufacturing and other interests of the Dominion; that such a policy . . . will encourage and develop an active reciprocity of Tariffs with our neighbours, so far as the varied interests of Canada may demand, will greatly tend to procure for this country, eventually, a reciprocity of Trade." So the idea was that tariffs would boost manufacturing to the point where other countries, especially the United States, would beat down Canada's doors to get favourable trade arrangements.

The National Policy was not without benefits. There is no doubt that the tariffs coincided with Canada's industrialization and the reversal of Canada's emigration problem. By 1900, population flows had stabilized and, in fact, the country began to grow dramatically. Manufacturing workers outnumbered farmers by the time of the 1911 census. An undeniable boom in American branch plant investment had created an American manufacturing sector in miniature. Canadian domestic industry was a mixed bag, but a few Canadian companies like farm implements manufacturer Massey-Harris emerged as strong, competitive global exporters.

The National Policy also created a stable and powerful political constituency for itself. Manufacturing interests successfully pitted industrial labourers and western farmers against each other. The real enemies of the manufacturers and their tariffs were the finance and railway interests, but they found ways to appropriate parts of the surplus for themselves through their monopolistic control of rates. With everyone who mattered more or less happy, the tariffs proved politically sustainable.

What were the costs? First was the simple fact of the tariffs themselves. They meant higher prices on a range of necessary goods, particularly finished manufactured goods, that hit hardest among the poor; in a very real sense, this was an upward transfer of wealth. The burden bordered on the absurd. The supposed success of Canadian farm-implement manufacturers meant that in the first decade of the twentieth century, Canadian wheat farmers on the Prairies paid about 15 percent more for farm implements than their neighbours in the American Midwest, allowing the Americans to break even at a lower price of wheat. Wheat farmers all over the world enjoyed more competitive conditions for themselves relative to their Canadian counterparts thanks to the National Policy.

There was also a subtle, long-term cost to Canada's development. We did not grow a domestic capital market to make sound and durable investments in industrial ventures. Other late industrializers such as Germany, Russia, and Japan developed various institutional mechanisms to direct industrial development. Canada instead relied on foreign direct investment and the patchwork efforts of Canadian industrial firms to secure long-term capital through debt or sales of equity. The foreign firms that set up in Canada had no ambition beyond extending their existing lines of business in the Canadian market. Their capacity to innovate remained in their home countries, along with their skilled technical personnel and research capacity.[8] The Canadian Manufacturers' Association bemoaned the lack of vocational, technical, and industrial research talent in Canada in their official magazine: "lawyers' offices are overstocked with impecunious new graduates, so too physicians' . . . It is this surplus in an honourable profession that supplies the demand for abortionists and similar questionable characters."

8   Two economic history doorstops were written in the late twentieth century: R.T. Naylor's *The History of Canadian Business, 1867-1914* and Michael Bliss' *Northern Enterprise*. Naylor comes at economic history from a nationalist, developmental lens and Bliss from a liberal business triumphalist perspective. There is not much that the two agree on, but both criticize the National Policy for creating unsustainable growth that left Canada's industrial sector a weaker copy of its American counterpart, both unable and unwilling to compete globally.

It didn't help matters that Canadian ventures were often shoddily run. One English visitor to Canada commented, politely, that Canadian eagerness to jump into business lines with which they were not well acquainted "does more credit to their courage than to their prudence."

The combination of inefficient domestic companies, insulated by the tariffs, and well-run American businesses content to take easy wins from the Canadian market did Canada no favours over the long term.

The lack of investment in technical education and industrial research was especially glaring as the old industries of the first Industrial Revolution, textiles and railroads, were supplanted in the last decades of the nineteenth century by the industries of the second industrial revolution, such as chemicals, petroleum, engineering, and tooling. These required serious capital investment for be success. The American economic historian Brad DeLong argues in *Slouching Towards Utopia* that the long twentieth century, a period characterized by unprecedented economic and technological expansion, began in 1870 with social innovation in the form of the modern industrial corporation and the research laboratory. Canada did not have any domestic industrial research capacity until the Great War.

A weak capital market together with a lack of capacity for cutting-edge products and techniques left a tragic gap in industrial development capital. The gap was filled by the most wasteful form of business subsidies: a ruinous game of competitive bidding among municipalities to attract businesses within their borders. This "bonusing" system led cities and towns to dig deep. Gifts of cash or land, loans, utility discounts, tax exemptions, wage subsidies, infrastructure and even dividend guarantees flowed feverishly from towns desperate to beggar themselves with giveaways before their neighbours could do the same. This was not just incredibly regressive, as poorer citizens shouldered the increased utility rates and taxes needed to cover the giveaways (well into the future, since municipalities often turned to issuing debt), but forced existing businesses to subsidize their own competitors. This bonanza of giveaways led to one newspaper drily observing that "banana growing in Manitoba could be made profitable on the same terms."

From the first railway investments, through the imposition of tariffs to the municipal bonusing craze, early Canadian government efforts to shape and stimulate the economy rarely worked out for Canadian working people. They were left with bills for the largesse granted to businesses. And on the eve of the Great War, their economy was top-heavy with overbuilt railroads, deficient in industrial capital, and dependent on uninventive foreign-owned manufacturers.

CHAPTER THREE

LEARNING ECONOMICS THE HARD WAY: RECESSION, DEPRESSION AND C.D. HOWE

After a half-century of industrial strategy culminating in an unbalanced and uncompetitive industrial economy, Canada entered a crisis period that spanned from the Great War until the end of the Second World War. The First World War initially benefited both business and government. Canadian industrial productivity surged as the economy responded to a flood of production orders through the Imperial Munitions Board. Between a quarter and a third of British artillery ammunition was made in Canada, and the dominion supported its imperial patron with bounteous food exports. Bond investors made out quite handsomely from war finance. The Wartime production effort required unprecedented levels of government involvement in the economy, although the structures that enabled this were gradually wound up after the war amid concerns around economic instability. The good times didn't last, and the legacies of the turmoil of the interwar years and Second World War were new currents of radicalism and strengthened tools of state intervention that whetted Canadians', and their governments', appetites for activist economic policies in pursuit of the common good.

After the Great War, inflation and a tight labour market gave way to deflation and depression. Trade in Canadian staples, which had been interwoven with British imperial capital and export markets, began to unravel. This was partially offset by increased American investment, which sextupled between 1914 and 1927. All the same, the coal and steel industries, mostly based in Cape Breton, collapsed, and 150,000 Maritimers left the country for the US in the 1920s. This led to Canada's first major experiment in regional development in an area not on the frontiers of settlement.

The government of William Lyon Mackenzie King, under heavy lobbying pressure from the Atlantic provinces and many of his own MPs, put together an aid package for the struggling industries. King turned down later requests for assistance and noted gloomily in his diary that "the whole problem seems a futile effort, to combat geographical and other economic considerations." Indeed, Maritimers were not alone in hitting the bricks for greener pastures: 300,000 Canadians left the country for the US in 1924 alone, and around 1 million did so over the course of the decade.

The 1920s, already challenging, ended on a disastrous note as the Great Depression struck, ensuring that the 1930s would be even more difficult. The politics of this new depression differed significantly from previous downturns. Canada's staid Victorian and Edwardian politics were increasingly disturbed by populist movements. Canadian radicalism had a distinctly agrarian bent: farmers' parties won provincial elections outright, and Thomas Crerar's agrarian Progressive Party won fifty-eight seats and over 20 percent of the vote in the 1921 federal election, beating out Arthur Meighen's Conservatives to become the official opposition. Alberta's weirdo-populist Social Credit Party, under William "Bible Bill" Aberhart, emerged from nowhere to form government after the 1934 provincial election, and pursued a variety of dubious monetary reforms. Labour activism was also on the rise, most notably with the 1919 Winnipeg General Strike. Smaller labour-led and more doctrinaire socialist parties emerged to serve as the parliamentary arms of the industrial proletariat. The Co-operative Commonwealth Federation was founded as an agrarian-labour socialist party in the early 1930s.

R. B. Bennett's Conservative government, elected in 1930, scrambled to respond to calls for relief from business and other orders of government, as well as a more demanding public. The populist and socialist agitation for greater degrees of government planning and economic intervention in the economy to some extent forced his hand. As David Lewis, then a significant figure in the CCF and later leader of the federal New Democratic Party, said during the war, many Canadians radicalized by the Great Depression "began learning economics the hard way."

Bennett's Canadian New Deal raised tariffs, established various marketing boards, and provided ongoing support to the coal and steel industries and the railways, but these measures had limited success. Big business hated them. Many of Bennett's measures were savaged in the courts. Voters kicked Bennett's government to the curb in 1935 and replaced it, once more, with Mackenzie King's Liberals. The Liberal government's National Employment Commission made far-reaching recommendations on Keynesian lines–stimulative public spending on public works to get the economy back into gear–in 1938. These received a warm response even from the business groups that had earlier resisted Bennett's New Deal measures. This period also saw the rise of Clarence Decatur (C.D.) Howe, an American-born engineer who would become one of the most significant figures in Canadian economic history.

Michael Bliss' magisterial *Northern Enterprise* calls its chapter on Howe, whom he clearly does not like, "The Years of C.D. Howe," perhaps in homage to Robert Caro's even more magisterial *The Years of Lyndon Johnson.* Howe fits the mould for multi-volume meditations on the nature and use of power: a peculiarly mid-century, larger-than-life, psychotically autocratic, workaholic figure very much like Caro's muses, Robert Moses and Lyndon Johnson. There was a certain class of these men in business and politics who by virtue of a confluence of ability, happenstance, and the alignment of political and economic change outgrew the nineteenth-century world into which they had been born—a world of globalized trade and finance, the gold standard, and the night watchman state—and became midwives to a peculiarly North American brand of big-government liberalism.

Howe had cut his teeth in engineering and contracting. His firm built most Western Canadian terminal grain elevators. The Liberal Party had courted this successful businessman to run in the new Northern Ontario seat of Port Arthur (now part of Thunder Bay) in 1935. When the Liberals were swept back to power, the newly elected Howe was named minister of railways and canals. He quickly merged these files with the Marine Department to create the Transportation Department. His stint in Transportation led to an abiding interest in civil aviation and the creation of Trans-Canada Air Lines (TCAL) in 1937, an entity controlled by the federal government through the nationalized Canadian National Railways.

Howe's airline faced opposition from the Canadian Pacific Railway and its executives. They had been invited to participate as a junior partner in the creation of TCAL, but shied away on the grounds that their involvement would amount to a *de facto* endorsement of state capitalism. Howe, with his engineering background and managerial, efficiency-seeking mindset, did not share their horror of state participation in the economy.

The Second World War began in September 1939. That month proved lucrative for Ottawa's stately Château Laurier: more cigars and champagne were sold than in the entire previous year, as the war industries geared up to receive lucrative contracts. Howe was appointed minister of munitions and supply at the beginning of 1940, and on April 9, the same day that the Nazis invaded Denmark and Norway, legislation creating a department of the same name came into force.

Howe's efforts to organize and often direct wartime production resulted in the creation of twenty-eight crown corporations to lead where capital would not, including, for example, the creation of a synthetic rubber plant in Sarnia. Another important trend in Canada's wartime economy was integration with the United States. Where the Great War had seen Canada act in lockstep with Britain, Howe was keen to pursue a continental approach to war production. Of course, as an American by birth, Howe had less Anglophile nostalgia for the empire than many of his contemporaries.

Wartime regulation of profits chafed big business interests, despite the sweetener of federal loans, grants, and tax write-offs. Algoma Steel received $20 million to modernize its plant, and the Aluminum Company of Canada was able to write off $155 million of a $237 million investment. As Bliss notes, many other companies "emerged from the war with enlarged, modern facilities fully paid for from untaxed proceeds of war work."

Even with that, Howe's relationship with Canada's industrial barons was in a glacial chill when he went to Britain at the end of 1940 to coordinate production with the British government. During his absence, his domestic enemies put pressure on Mackenzie King to get rid of him. Howe survived both this attempt and the torpedoing of his ship 300 miles from the coast of Iceland. He returned to Ottawa to fulminate in the House of Commons against the big-business perpetrators of the attempted behind-the-scenes coup against him. "The number one saboteur since the beginning of the war is the *Financial Post,*" he declared, a reference to the preferred newspaper of Canadian capitalists. The notion of a coup was exaggerated, but the business community was indeed displeased with what it saw as Howe's autocratic, overreaching manner. After this episode, relations with business largely normalized. The barons recognized that the determined Howe was there to stay.

At the same time that Howe was busy coordinating war production with his mix of carrots, sticks, and Crown corporations, the federal government's Department of Finance was struggling to figure out how to pay for the war effort.

During the Great War, Canada had implemented its first direct federal income tax. Borden's finance minister, Sir Thomas White, had resisted an income tax for years: he felt it was unfair to duplicate provincial and municipal tax burdens, and that it would adversely affect Canada's tax competitiveness with the United States while also presenting a considerable administrative headache for a thinly-staffed federal government. His change of heart in 1917 was prompted at least in part by the need to extend credit for the purchase of $40 million worth of cheese for export to Britain.

The 1917 Income tax was so low as to be largely irrelevant to Canada's war finance efforts (and after the peace, it dwindled throughout the 1920s due to political pressure). The government experimented with other revenue-generating alternatives during the Great War. There was an excess profits tax in 1916. A land tax and outright wealth confiscation (a capital levy) were considered and rejected by the government as overreaches. The marquee financial innovation, in terms of actual financial impact, was the war bond, marketed and sold to the general public on generous terms. Payments from Victory Loans and a few other domestic debt issues were tax-free. While public response exceeded expectations, big investors were able to make the most of this tax advantage.

Through the course of the 1920s, regressive taxes such as tariffs and sales taxes, which hit all individuals equally regardless of income, gained ground at the expense of more progressive personal and corporate income taxes, which better reflected the taxpayer's ability to pay. Ordinary Canadians were being squeezed while big investors in government debt enjoyed tax-free profits.

The perception that the big financial winners of the Great War were rich Institutional Investors informed how the government would approach finance in the next conflict. At the start of the Second World War, there were approximately 300,000 Canadians paying a federal income tax. In 1940, Finance Minister J.L. Ilsley and his team at Finance, including Clifford Clark, perhaps Canada's best-known deputy minister of Finance (I concede this is not a high bar), introduced a new National Defence Tax (NDT). By 1941, the personal income tax and the NDT combined to boost the number of federal income taxpayers to 870,000 out of a labour force of 3.5 million, nearly triple the pre-war level. This rapidly expanded to 2.1 million taxpayers in 1943, after the 1942 budget harmonized the two taxes as "graduated" and "normal" taxes. The number rose to 2.7 million by 1951.

The new tax structure included some claw-back points where earners began to earn less as they approached the minimum exempt amount, and as Ilsley noted in Cabinet, "the small man is complaining bitterly for the first time." Countless letters poured into Finance and the department seems to have taken them seriously. Deputy Minister Clark directed senior staff to respond personally to many of them, and also introduce marginal tweaks to make the tax more equitable to low-income earners.

Despite the large increase in the number of taxpayers and the much more balanced blend in revenue sources in paying for the war (a roughly even split between borrowing and taxes), the contributions of working-class Canadians were fairly marginal to the government's overall revenue picture. Tax historian Shirley Tillotson argues that the incidence of taxes on poorer Canadians was important precisely because it was marginal: it was a "show them the price" tax, "more about preventing the poor from being free-riders and making them see themselves as tax-conscious taxpayers than it was about buying bombers and feeding the troops." This light touch on less wealthy Canadians would not survive the coming tide of tax allowances and credits for business that formed the backbone of the post-war corporate welfare state.

To return to our friend C. D. Howe, he had been appointed minister of reconstruction in 1944, with eventual Allied victory in sight. In a speech to the Reform Club of Montreal in November 1943, he had said that the "problems of peace could be solved by the solutions used during the war." Following the UK's popular Beveridge Report on social welfare and reconstruction, Prime Minister Mackenzie King commissioned a similar report on social security measures. Mackenzie King's party was keenly aware that the CCF was riding high—dangerously so, from a Liberal perspective—in public opinion polling during the later years of the war, and that the 1944 election could lend its opponent an opportunity for a decisive breakthrough and a different vision of economic management. David Lewis had written in a popular 1943 book that "the question is, therefore, no longer whether we shall have a plan or not. The problem is rather whether we shall have planning for monopoly by monopoly or whether we shall build a co-operative organization of industry to make possible planning by the people for the people." The federal government's experiments with major interventions in the economy during the war had left Canadians confident that their government was well equipped for an ambitious program of government-led economic and social reconstruction in the post-war era–and the Liberals had to respond to that demand.

The Liberal response to the increasing popularity of plans for peace-time prosperity was to expand collective bargaining, introduce family allowances, launch a successful and ambitious program of homebuilding, and articulate a commitment to the Keynesian notion of using fiscal and monetary policies to smooth fluctuations in the business cycle. The latter step was revealed by Howe himself in his new capacity as minister of reconstruction. It was welcomed by important members of the business community. The chairman of the Canadian Chamber of Commerce conceded in 1943 that business leaders had been "wrong ten or twelve years ago in their approach to government policies and fiscal policies and we are now taking exactly the opposite position, and we hope that if government in the future tries to encourage business in the days of depression, and tries to repress business in boom days, business on the whole will understand that [it is] a pretty sensible thing for the government to do, and won't make it any harder for government than possible."

The government, it must be said, did not embrace a social-democratic approach to the economy, with widespread public ownership and planning, in the wake of the war. Canada developed a more fulsome welfare state that made life better for very many Canadians, and the government itself was significantly larger and more involved in the economy. But most government operations continued to follow established practice: subsidization of industry, with considerable preference and deference to business.

It is instructive to look at the agenda of Howe's Department of Reconstruction and Supply. The federal government produced excellent year books through the Dominion Bureau of Statistics (predecessor of today's Statistics Canada) with administrative snapshots of the government's activity. The chapter dedicated to Howe's department in 1945 noted that "the first early task of reconstruction will be to facilitate and encourage an expansion of private industry," with a focus on standing up an Industrial Development Bank, industrial reconversion and sale of surplus assets, industrial research (today's Industrial Research Assistance Program), and working through the claims of depreciation allowances—tax write-offs for industrial investments—by businesses. Howe was often referred to as the government's "minister of everything," and a look at his department's remit in 1946 supports this. It included "industrial development and conversion; public works and improvements; housing and community planning; research and the conservation and development of natural resources."

Depreciation allowances, which had been extended several times during the war, allowed companies to claim depreciation at twice the normal rate against their taxes. The point of depreciation allowances was to maintain a rate of return on wartime investment that was acceptable to capital in light of formal profit controls. In the end, this cost the federal government $1.4 billion in foregone revenue—at a time, it should be repeated, when many working Canadians were first taking on a real portion of the country's direct tax burden. Business defended the policy by pointing to the UK's acceleration of depreciation allowances, overlooking that British wartime and post-war measures existed in a context where industry had been pulverized by German bombs. Canada, of course, was spared widespread physical devastation during the war. Government sided with business. Capital owners were granted this public support in the reconstruction era. As the CCF's David Lewis put it in his wartime book *Make This Your Canada*, "the unpatriotic strike of capital for higher profit was accepted as a matter of course, in spite of the outbreak of war. Such are the rules of the capitalist game that only a strike of workers for higher wages is called unpatriotic." Theoretically, requests for accelerated depreciation were now to be adjudicated by the government. This was usually done on a gentlemanly basis. About 80 percent of capital investment in manufacturing between 1945-49 qualified, despite the war being over for the vast majority of that period.

Louis St. Laurent's government made accelerated depreciation tax breaks permanent in the post-war years. John Diefenbaker's Progressive Conservative government allowed time-limited and region-limited double depreciation allowances in the early 1960s, adding some nationalist and regional development elements to the general policy of subsidy for investment. Lester Pearson continued on the nationalist track by favouring companies that were at least 25 percent Canadian-owned or controlled, and Pierre Trudeau later added generous allowances for certain sectors and kinds of investment.

* * *

While these changes to the tax code largely aimed to attract and stimulate industrial investment, especially in manufacturing and mining, governments also began to systematically subsidize companies' research and development in an attempt to make Canada's economy more productive and innovative. There had been a provision tucked into the Income Tax Act as early as 1944 that allowed companies to deduct their research expenditures, including a partial deduction of capital expenditures related to research. The Diefenbaker government expanded this with Canada's first special tax incentives for R&D in 1962, a foundation upon which Pearson's government built with the Industrial Research and Development Act in 1967, replacing the tax credit with cash grants.

The post-war years also saw the expansion of direct subsidies in pursuit of regional development and industrial excellence. The St. Laurent government continued to heavily subsidize the Cape Breton coal and steel industries and jumped headlong into the disastrous Avro Arrow jet fighter project—perhaps Canada's most well-known failed defence procurement, but certainly not its last.

The Diefenbaker government, elected on a mildly economic nationalist platform—then a fashionable position in the wake of the high-profile Gordon Commission's report on Canada's economic prospects—aimed at boosting the prospects of Canadian manufacturing through targeted subsidy policies in the late 1950s, as well as a variety of regional and rural development programs.

Lester Pearson built upon these with the Area Development Agency within his new Department of Industry and followed it up shortly with the Area Development Incentives Act (ADIA) in 1965, one of the precursors to the formal structure of regional development policy that Pierre Trudeau would expand upon with the creation of the Department of Regional Economic Expansion (DREE) in 1969. Pearson's government also created new technology-oriented subsidy granting programs, most notably the Program for the Advancement of Industrial Technology (PAIT) in 1965, which was found by an evaluation to suffer from inconsistent and

41

unfocused objectives and an over-focus on projects that were already likely to be successful.

Access to cheap capital was another element of the post-war toolkit, although much less significant than tax breaks and direct subsidies. Howe's new Industrial Development Bank (IDB) had been intended to supplement Canada's pool of industrial capital. Wartime meetings of the House of Commons' banking and commerce committee estimated the need for $1.5 billion annually in capital expenditure to maintain full employment, but a survey of over 2,000 businesses learned that only $106 million in private capital expenditure was expected. As we saw in the last chapter, Canada had developed no native institutions specialized in industrial finance. By 1944, long-term finance, including industrial loans and mortgages, made up only $154 million of nearly $6 billion of total bank assets—around 2.5 percent of the total. There was a clear need to supplement the private sector. But institutional guardrails to ensure that the IDB did not compete with established banks also ensured that it would remain a small and peripheral part of Canada's capital market. The IDB could have been a genuinely useful addition to Canada's suite of industrial and employment policy measures, combining operational independence with a long-term policy vision, but Mackenzie King's Liberals were unwilling to pick a fight with finance. The IDB's focus shifted over time to small business loans, and the government would eventually change its name to the Business Development Bank of Canada, shortened to BDC in the 1990s.

None of this would have been possible without dedicated machinery inside of government. Most of the Crown corporations stood up during the war had been liquidated at fire-sale prices of about 35 cents on the dollar. In some fairness, most were producers of war materiel with little relevance to the economy outside of wartime mobilization. But Howe did maintain a few Crown corps within the federal family, including the Eldorado uranium mine, a crown jewel for the atomic age, and Sarnia's synthetic rubber plant. He remained willing to use his department, which maintained a sophisticated economic research branch that followed Howe when he moved over to Trade and Commerce, to identify and support the industries of the future, including his lifelong object of affection, aviation. Later governments built on this. Lester Pearson's government created in 1963 the Department of Industry to be "for the manufacturing industry what the Department of Agriculture is for farmers." Walter Gordon and other economic nationalists around Pearson strongly influenced its creation and the department carried that DNA in its bones.

At the same time that many of the core mechanisms of the next generation of federal business subsidies were clicking into place—tax allowances, subsidized credit and capital, export and research assistance, and selective support for industries—provincial governments, too, were getting in on the act. Before the war, only four provinces had departments focused on industrial development: Quebec (Trade and Commerce), Alberta (Industries and Labour), BC (Trade and Industries), and Nova Scotia (Trade and Industry). Manitoba may count as a fifth, with a publicly supported Industrial Development Board. All were founded during the 1930s. After the war, they were joined by Ontario (Planning and Development), New Brunswick (Industry and Reconstruction), Saskatchewan (Natural Resources and Industrial Development as well as Reconstruction and Rehabilitation), and formally this time, Manitoba (Industry and Commerce within Mines and Natural Resources). Only Prince Edward Island failed to create one. Provincial governments from coast to coast were keen to use the new fiscal tools to spur industrial and resource development and deliver on the reconstruction promise of post-war abundance. Provincial governments, unfortunately, tended toward dramatic, politically inspired megaprojects or big bets based around personalized dealmaking, most of which crashed and burned.

* * *

Canada's newest province provided an instructive example. In 1949, after a contentious two-round referendum, the colony of Newfoundland and Labrador opted to join Confederation as Canada's tenth province. The brutal experience of the Depression had brought the Dominion of Newfoundland to financial ruin and led to a return to direct colonial administration under the imperial Commission of Government. After the war, the colony voted (narrowly) to join Canada as a province, and soon elected a Liberal government led by the delightfully odd Joey Smallwood.

Smallwood's government was immediately beset by crisis. The grand bargain of Newfoundland joining Confederation—a higher standard of living delivered in part through new federal programs—came with the trade-off of Newfoundland eliminating its tariffs. The province's industries promptly faced hot competition from the mainland that threatened to drive many out of business. The 1949 British currency crisis and devaluation of the pound made Newfoundland's exports, especially of cod and timber, less competitive. American military bases on the island had been shuttered, and these closures and the downturns in fishing and lumber combined to create massive unemployment. The constant trickle of outmigration to the mainland threatened to turn into a flood now that Newfoundlanders and Canadians were within the same federal family.

Smallwood's vision on leading the colony into Confederation had been that rising standards of living would decimate fishing outports and traditional industries, necessitating a new economic, cultural and political self-confidence in Newfoundland, something that only modern industrialism could bring. Newfoundland's post-Confederation crisis sharpened his belief. "Develop or perish" became his motto, and after early attempts to set up an Industrial Development Loan Board, modelled on the federal Industrial Development Bank, failed to produce quick results, Smallwood felt that he had to gamble on economic development a chunk of the $40 million or so in surpluses accumulated during the Commission of Government period. He gladly conceded that he was rolling the dice: "People are not going to wait forever for this development; if we don't give it to them tomorrow, they . . . go where jobs are, you cannot blame them. Our job is to back them, go right out, boots and all, make or break. Here, what I mean by 'make or break,' here is gamble."

On a visit to Ottawa, he asked C.D. Howe to recommend someone who could manage a new Department of Economic Development. Howe recommended Alfred Valdmanis, a Lithuanian I who had served as his country's Minister of Finance before the war and had led a relatively successful effort to diversify and industrialize the small, agrarian Baltic nation. Valdmanis was hired more or less on the spot in May 1950.

Thus began a genuinely bizarre two-year spree to set up as many new industrial operations in Newfoundland and Labrador as could possibly be managed. Valdmanis leveraged European contacts to set up a cement factory in Corner Brook. When West German capital controls prevented the German corporation from making cash contributions, and American investors insisted on delaying to conduct their own analysis, Smallwood opted to build the factory at provincial expense in August 1950. A gypsum plant and hardwood processor soon followed. Screening and analysis grew sloppier and Valdmanis' claims grew more outlandish, promising thirteen new industries in 1951 and the effective elimination of unemployment in 1952. This Rasputin of the Rock thoroughly alienated cabinet members and Smallwood's personal friends, but no matter. After West German capital controls were eliminated in January 1952, investment poured in and was matched on a fictional "dollar-for-dollar" basis—the province counted "in kind" contributions of expertise and machinery as "dollars" and advanced the majority of the cash required by the projects.

The federal government began to worry about Smallwood's hectic pace. Delays and cost overruns piled up, and Valdmanis' nerve eventually broke. He asked Smallwood to suspend the investment attraction program in favour of a period of consolidation. He offered his resignation as Director General of Economic Development in 1953 and the next year resigned from the board of the province's development and resources holding company, NALCO, under fire from Smallwood for opening an expensive office in Montreal. He was arrested, prosecuted and convicted of fraud, having received kickbacks from some of the German companies he enticed to Newfoundland.

Smallwood's industrialization blitz had been heavily focused on personalized dealmaking, a fairly aggressive propaganda campaign, and a noted lack of careful analysis or strategy, especially as time went on. The early cement and gypsum factories, at least, had been selected based on availability of natural resources and a strong local market amid the postwar housing boom. These were eventually sold at a profit. But any industry was a good industry to Smallwood, and Newfoundland ran through the Commission-era surplus without much to show for it. This was crude post-colonial economic nationalism, rendered tragic by the imperative to act created by Newfoundland's genuinely crushing post-Confederation economic impasse. By the time the dust had settled, observers felt it fair to assert that "had Smallwood literally taken $10-15 million and burned it in a bonfire atop Signal Hill, the end result, in terms of Newfoundland's progress, would not have been greatly different." If this was the tragedy of economic development efforts in Newfoundland, it would be succeeded in due time by a series of farces.[9]

This was the end result of the experience of war finance and economic planning in Canada, as well as the rise of radical and populist threats to the established order during the Depression and Second World War. They led to the construction of a rudimentary welfare state and an embrace of a relatively passive[10] style of government economic management via stimulating demand and investment through tax breaks, credit, and subsidies. In the heady environment of victory, the federal and provincial governments had been ready to use the tools of the war to build the peace. The approaches they chose, however, amounted mostly to subsidies for private business to do the job for them. Budgets and monetary policy were managed to keep employment high and inflation controlled. Confident bets were laid on subsidies to incentivize private investments to launch frontier industries, maintain jobs in mature sectors, and rejuvenate and develop poor regions. As subsidies piled upon subsidies during the postwar decades, Canada entered its golden age of corporate welfare.

---

9 Brian Peckford's Progressive Conservative government attempted to launch a greenhouse-grown cucumber industry in the late 1980s that ruined the government's credibility, and the attempts of every provincial government from Danny Williams' on to manage the fiasco of the Muskrat Falls hydroelectric project certainly qualifies, though perhaps as a tragic farce.

10 Compared to economic planning.

CHAPTER FOUR

CORPORATE WELFARE BUMS: THE GOLDEN AGE OF HANDOUTS

The decades immediately after the war were good for Canada. Economic growth was steady and the population expanded by 54 percent from 1951 to 1971. The Canadian economy became more tightly integrated with that of the United States through both American foreign direct investment and deliberate policies such as the 1956 Defence Production Sharing Agreement and the 1965 Auto Pact. The economic importance of old imperial connections to Britain declined along with Britain itself. Despite broad and sustained growth, Canada's secondary exporting industries lost ground, particularly in high-tech sectors.

By the federal election of 1972, subsidies to businesses had proliferated dramatically. Liberal Prime Minister Pierre Trudeau sought a renewed majority government after his Trudeaumania-era romp to victory in 1968. Robert Stanfield gamely stepped back up to the plate, leading the Progressive Conservatives. Our friend, David Lewis, was by now leader of the New Democratic Party. The explosion of business subsidies over the prior decades played an important role in the race, with the NDP and Lewis taking aim on the campaign trail at "corporate welfare bums." Lewis' NDP managed to make small gains and, crucially, for the first time in its history, secured the balance of power in the House of Commons as the only party absent the official opposition with whom the Liberal minority could pass a budget.

Lewis, who had won the party's leadership as a relative moderate in 1971 against a fierce left-wing challenge from the radical Waffle movement, published a short but detailed book for the campaign that expanded on his arguments against corporate welfare. Lewis's book points to the invigoration of the system of Canadian corporate welfare in the moment of post-war reconstruction we have just discussed: left with the prospect of a sharp economic contraction after the end of the war, Ottawa made a choice. Instead of embarking on a program of stimulating consumption through the provision of social goods and managing investment policies more directly, as Lewis would have preferred, the federal government opted to stimulate private investment. The federal government steadily built and expanded upon a suite of tools to do this: tax measures, especially the accelerated depreciation allowances,[11] but also cheap credit, research subsidies and the sale of government assets for pennies on the dollar.

Lewis' analysis of subsidy programs across government, leaving aside the political accusation of Liberal fecklessness and coziness with big business, was that they suffered from aimlessness. He excoriated the federal government's generations-long subsidization campaign of the shipbuilding industry "on an on-again, off-again basis, never with a carefully thought-out, long-range plan in mind," to say nothing of the emerging landscape of government innovation and research support programming designed to stimulate economic growth, where each program seems to operate in isolation from all the others, and each has little to show for its efforts.

Modern readers from all across the political spectrum would sympathize with Lewis' palpable anger at the tax giveaways and opaque programming that contributed almost nothing to lasting prosperity—"one ad hoc policy was piled on top of another" and "each new government, and almost every new cabinet minister, added to the list of giveaway programs, oblivious to what had been done before and probably unaware of what others were doing."

---

11 A point with which Michael Bliss would agree: his view was that depreciation allowances and other tax measures "evolved from a badly understood pump-priming device into the chief tool of industrial policy-making."

Business-friendly tax measures were singled out by Lewis as particularly obnoxious. The governments of Louis St. Laurent, John Diefenbaker, Lester Pearson, and Pierre Trudeau had all added their own touches to the tax code, which by 1972 had grown dramatically from the end of the war. The code grew so complex that in 1969, the dominion statistician had to announce the replacement of the existing annual report, covering corporate taxation revenues, with a series of specific reports, noting that "corporation taxable income has been affected by the industry in which the firm operates, the scope of the firm's operations, the geographical location, the ownership of the firm and several other factors. As a result, it was becoming increasingly difficult to satisfy the needs for data of both corporation finance generally and corporation income taxation." Over the decades, as the tax code became more generous to companies, Lewis observed the share of corporate taxes had declined precipitously from 28 percent of revenues (in 1951) to 12.2 percent in the 1972 budget. Over the same period, the share taken from personal income taxes nearly doubled, from 26.7 percent to 49.9 percent. Canadians, in a time where giveaways to business were reaching new heights, were shouldering more and more of the tax burden themselves.

These research tax credit schemes would only get worse in the years after Lewis published his book. The Trudeau government would eventually launch its Scientific Research Tax Credit in 1983. Companies eligible for the credit could issue securities that essentially transferred the amount of the credit to another company, creating a financial product that allowed companies conducting research activities to book profits just from the sale of the credits. The cost of the credit had been anticipated at $100-200 million, but these "quick flip" transactions ran the tab up to $2 billion before the loophole was hastily shut the next year and the credit replaced with the modern Scientific Research and Experimental Development Tax Credit (SR&ED) a few years later.

That story, while amusing and frustrating in equal parts, points to an inherent weakness of tax credits as tools to solve an economy's ills. It is extremely difficult to tune a credit such that it is at once accessible and efficient—that adequately balances the administrative burden and general headache for eligible recipients with ensuring that credited dollars actually incentivize what the credit is meant to promote. Too accessible, and the credit becomes an open-ended fiscal commitment to subsidize business generally; too targeted and only the biggest companies with the best tax accountants will avail themselves of it.

In the case of research tax credits, smaller research-focused companies have to consider if the considerable and expensive work involved in demonstrating their research and development *bona fides* is worth the money. This is a perennial and ongoing problem that has spawned a whole industry of compliance consultants, not to mention the technology company equivalent of payday lenders who lend at high rates against a company's future SR&ED credits. On the other side are big companies, who have no difficulty claiming credits and juicing them for all they are worth, but whose economic gains may well be realized outside of Canada.

While David Lewis' little campaign book's strongest ire was for these tax giveaways, it also took aim at direct subsidy policies as designed and implemented by the governments of the day. Here, too, governments had built extensively on the post-war moment. Subsidies tended to have one of two outcomes in mind. The first was stimulating investment in industries or sectors that the government saw as strategic for geopolitical or economic reasons, essentially an attempt to "crowd in." The second was equity: that a region or group deserved "catch up" investment to counter for historic underdevelopment or lack of access to capital.

Regional development programs within DREE, growing out of rural anti-poverty programs, were primarily built on the equity mode. They became the first standalone institutional home for the long, rich Canadian practice of local pork-barrel patronage. Early handouts in 1971 and 1972, as Lewis noted, included subsidies to the building of new pulp and paper mills in Alberta and Quebec—this occurred during a weak period in the pulp market and drove an existing profitable firm in the rural Quebec region of Temiscamingue out of business, leaving 875 people unemployed; there were also layoffs at other existing pulp operations.

Easy money from government also lent itself to decision-making on a crassly political basis. More recent analysis by Kevin Milligan and Michael

Smart of the track records of regional development agency funding has found that government-held ridings received an extra $10 per capita in regional development spending, about 20 percent over the mean. Cabinet ministers' ridings did even better, with an extra $16.30 per person.

Lewis was scathing about DREE, despite his basic sympathy for the objective of assistance to struggling regions. The Liberals, he felt, just could not help themselves from turning it into both a feather-bedding exercise for MPs and Ministers, and a spigot from which big business could lap at its pleasure—Lewis noted that one of the government's industrial advisory groups was made up of executives from big companies who all ended up receiving grants themselves.

As for subsidies designed to counter downturns in the economy or stimulate investment in promising industries or sectors, Lewis noticed that they were sustained by perverse logic. "The tendency to escalate welfare handouts," he wrote, "gains momentum with each new failure in the economy." And even in good times, recipients of corporate welfare wanted more. Much as with Macdonald's National Policy, defenders of the system claimed that "infant industries needed tariff food to nourish them when healthy, and tariff medicine to protect them when sick." There was never a time when more public funding was a bad idea. The full spectrum of business subsidies took on an entirely unhealthy, self-reinforcing cast.

The business community was fine with it, on the whole. Companies were more concerned with scooping up funds than criticizing the government's profligacy. "Canadian businesses," wrote Lewis, "whatever their public pronouncements on the matter, not only acquiesce to government involvement in the economy but have come to depend on it." The recipients of handouts might have recognized the patent absurdity of subsidy proliferation, but they knew that if they did not put their hand out, someone else would. Near the end of his book, Lewis identified the iron logic that has long protected regional development and other corporate welfare programs with a delightful anecdote. A campaigning Prime Minister Trudeau, he writes, told a Halifax audience that "if people don't like [DREE grants], we'll give them all to Quebec."

Lewis's attack on corporate welfare bums was published just as the era of postwar social democracy in the North Atlantic world was coming to a close in the early 1970s. The global Bretton Woods monetary system folded. The first of the 1970s' global oil shocks hit, representing the beginning of the end of an era of high growth, high profits, and high employment. Governments suddenly were in uncharted economic waters and grappling with a new, much more difficult reality.

In the wake of the oil shocks and the brutal steps taken to tame stubborn inflation at the end of the 1970s, governments had little choice but to restrain themselves on handouts. But over the 1980s, the neoliberal revolution, brought about by the likes of Margaret Thatcher, Ronald Reagan, and Brian Mulroney, did not hesitate to reinvigorate old practices with generous policies of business support. Indeed, the desperate circumstances of oil price shocks, spiralling inflation, and attendant unemployment made some governments even keener to launch reckless economic gambles in the form of corporate handouts and subsidies.

Churning policies and departments further complicated the design and implementation of business subsidy policies through to the 1980s. This was not a new problem. In the early years of the Trudeau government, Industry Trade & Commerce had been known as "one of the most unwieldy and unstable bureaucracies in Ottawa," with constant reorganizations and infighting among those officials dedicated to promoting orthodox trade policies and others keen on an activist industrial agenda, all of it complicated by industry lobbyists asserting their own perspectives and priorities. Policy escaped any meaningful oversight or coordination as business subsidies proliferated. DREE faced a policy review in 1972 after the city of Montreal was designated one of the priority "special areas" of the supposedly regional program, an obviously politically indefensible decision as far as the rest of the country was concerned. The review led to decentralization of DREE's operations and expansion of its geographic scope throughout the country: over $3 billion was transferred directly to the provinces to support economic development via federal/provincial agreements from 1974 to 1981.

In 1982, amid a sharp recession, the government once again revisited the logic of regional economic development programming. Trudeau merged DREE with the Department of Industry, Trade and Commerce (IT&C) and changed its name to the Department of Regional Industrial Expansion (DRIE), and combined various service offerings into the supersized Industrial and Regional Development Program (IRDP), which assigned every census district to one of four tiers of development with an accompanying level of government support for investments. This move re-centralized economic development policy back within the federal government, partly to re-establish the visibility of the federal government amid the deep recession of the early 1980s.

At the same time, the Trudeau government created a new Ministry of State for Economic and Regional Development (MSERD) to act as a quasi-central agency in support of a powerful new Cabinet committee, the Board of Economic Development Ministers. The committee aimed to rationalize an unwieldy and ineffective system of governance and control of subsidy and development programs at the ministerial level. MSERD was also to continue administering the federal/provincial agreements that DREE had managed.

The Mulroney government did its own re-examination of the whole structure of regional economic development programs in 1986. They were duly re-decentralized out of the department into agencies, beginning with the Atlantic Canada Opportunities Agency in 1987. The Department of Industry, shorn of its regional and trade responsibilities, was left standing with science and technology policy at the eve of the Chretien era.

Repeated changes in focus and direction in an area of policymaking are usually signs of trouble. Indeed, Ottawa had enormous difficulty separating its stimulative policies meant to promote innovation and investment from its policies of regional economic development. The institutional machinery meant to shepherd these priorities along was merged, split, and reconstituted with abandon, usually in response to short-run political pressures, right into the nineties. None of the shuffling led to a demonstrable improvement in performance. Canada never developed an analogue to institutions like the Japanese Ministry of International Trade and Industry, which oversaw that country's stunning postwar rebound, to serve as a respected, expert body managing economic development and industrial policy.

* * *

The provinces, too, continued to chase dramatic economic transformations to little avail through the seventies and eighties. In 1973, American car impresario Malcolm Bricklin announced that he would manufacture a futuristic new sports car, the Bricklin SV1, complete with gull-wing doors and advanced safety features. Unlike every other car assembled and built in North America, however, this one would be made in New Brunswick, thanks to a deal struck between Bricklin and Richard Hatfield's new Progressive Conservative provincial government.

Bricklin previously had been shopping his car to Quebec's government, which was skeptical of the plan's viability, leaving the door open to Hatfield. His province forked over $20 million and the federal government, through DREE, chipped in a few million more. For a poor, largely rural province characterized by often-caustic linguistic and cultural division, Hatfield saw the futuristic Bricklin, designed by the same stylist who designed the Batmobile in the 1960s Adam West Batman TV show, as a source of jobs and prosperity, and a signal that New Brunswick would finally develop a modern economy. It did not seem totally crazy to Hatfield that New Brunswick could take this leap—Nova Scotia had successfully attracted a Volvo plant and assembly plants for Japanese autos in the late sixties under Robert Stanfield's Progressive Conservative government.

Hatfield attended the Bricklin plant's opening and declared that the company's "success will be our success. We have come this far despite the pessimism of some politicians and the skepticism of self-appointed experts . . . We're building a better New Brunswick. If it were produced in Ontario or Quebec, there would be more optimism. We are not supposed to have dreams. This ceremony proves they are wrong."

Early reviews in the automotive press were positive. *Car and Driver*'s September 1973 cover asked readers who made the better sports car, Corvette or Bricklin. The answer, the magazine coyly noted, "may surprise you." Hatfield spent the 1974 election campaign crisscrossing the province in a Bricklin, and the PCs' campaign ads featured the car and its assembly line. Initial sales of the car were strong, but it was plagued by quality issues, and it was cripplingly expensive to make—each unit was sold at a loss. By fall 1975, the provincial government, which by that point owned 67 percent of the company, told Bricklin to find more money or it would wind up operations. The government soon followed through on that threat, leaving New Brunswick with nothing but a sad reminder that hope and an abiding belief in an underdog province aren't enough to succeed in advanced manufacturing.

In Saskatchewan, Grant Devine's Progressive Conservatives defeated Allan Blakeney's New Democrats in 1982 to form the first-ever PC majority government in the province's history. With the exception of a seven-year Liberal interregnum in the 1960s, the CCF-NDP dynasty had governed Saskatchewan from 1944. The CCF-NDP premiers, Tommy Douglas, Woodrow Lloyd, and Allan Blakeney, had been for the most part prudent slide-rule socialists who carefully managed provincial budgets, created new Crown corporations to exploit Saskatchewan's natural resources, and gradually expanded the province's welfare state, aided by a capable civil service. Devine's PCs saw his predecessors' careful fiscal policies and state-driven economic development strategies as a recipe for stagnation, high unemployment, and the slow death of the rural communities that were the province's backbone in the twentieth century. Hostile to the province's public service and convinced of the transformative power of deals, Devine's government made big investments in rural infrastructure and energy megaprojects.

In 1985, the government announced a deal with Federated Co-operatives Limited (FCL) to build an upgrader on the site of FCL's Regina refinery. Devine had repeatedly told his ministers that the province needed a megaproject, and FCL saw him coming. With its premier completely dedicated to getting to yes, the province repeatedly sabotaged its own negotiation team. FCL insisted on guarantees to be made whole in the event of losses, to own and operate the upgrader itself instead of operating it as a joint venture with joint oversight, and for the government to cover building costs. FCL understood that the upgrader would be a key element in Devine's 1986 re-election platform, and in the end, it got its way on all three points. Saskatchewan ponied up $158 million in cash, and

$637 million in loan guarantees (with an assist from Brian Mulroney's federal government).

The upgrader lost money In Its first three years. Saskatchewan, on the hook, had to cover the $75 million shortfall or allow a default and pay $400 million. Devine was meanwhile pursuing another upgrader project in Lloydminster with oil company Husky and the Alberta government. The provincial governments sunk hundreds of millions into that upgrader and were quickly left with a project that was hundreds of millions of dollars over budget. As Alberta Premier Ralph Klein later said of the Lloydminster upgrader, "We were all stupid together."

A general pattern of impulsive policymaking and loss of control over the budgeting process throughout the life of the Devine government ran Saskatchewan's books off a cliff. When Roy Romanow's NDP took over in 1991, the province was on the brink of default, leaving the new social democratic government to make bruising cuts.

Saskatchewan and New Brunswick were not entirely the masters of their own economic destinies. Small changes in commodity prices on global markets are more consequential than policy levers available to governments, especially in smaller jurisdictions whose economies are heavily dependent on agriculture and natural resources. Commodity prices, however, are not up for re-election every few years. Governments are. So reckless boosterism tends to characterize provincial approaches to economic development, particularly in smaller, more resource- or agriculture-dependent provinces. As Janice MacKinnon, NDP Premier Roy Romanow's finance minister, later wrote, the "one lever that governments did have was to persuade companies to come . . . by offering public investment on terms too good to refuse." This lever has, unfortunately for Canadians, been too tempting for provincial governments of all partisan stripes in every region of the country.

The growth of federal industrial support, regional development and research and development programs, and tax breaks through the 1970s and 1980s, from David Lewis' campaign against the "corporate welfare bums" through the Mulroney years, was genuinely staggering. The bums and their benefactors in Ottawa and in provincial capitals across Canada, it turned out, had been just getting warmed up when Lewis published his book. By 1984, the Mulroney government's task force on programs found that business subsidies had proliferated to $4.5 billion in grants and contributions and another $7.7 in tax expenditures, with another $4.9 billion spent delivering services: a total of $17.1 billion, or $43 billion in 2022 dollars. Exclusive of government salaries, this amounted to about 3 percent of Canada's GDP.

CHAPTER FIVE

GIVING WITH BOTH HANDS: WHY CORPORATE WELFARE FAILED CANADIANS

The corporate welfare state managed to spend a lot of money between the end of the Second World War and the 1980s. Many sophisticated, well-meaning people had set their minds to solving these problems of productivity, development, and diversification, but saturation bombing the private sector with subsidies failed to solve Canada's fundamental economic problems. Our industrial productivity was still low compared to the United States or Germany (especially after the oil shock in 1973). Regional economies in Atlantic Canada and other rural areas across the country failed to rebound, and the national economy remained heavily weighted toward the boom-and-bust energy sector. The bewildering array of generous subsidy programs always added up to less than the sum of their parts and Canada's economic dynamics remained unchanged. How did the giveaways so completely fail?

Over the years, different examinations of these problems have arrived at a consistent set of issues, led by the federal government's inability to coherently plan in both the economic and administrative senses of the term.

The question of how Ottawa spent was top of mind for Prime Minister Brian Mulroney when his new government took office in 1984. One of his first acts was to strike a ministerial task force on program review headed by the old Progressive Conservative hand, now deputy prime minister, Erik Nielsen. It was Nielsen's job to rationalize the thicket of government subsidy programs that had grown, more or less uncontrolled, since the war.

An introductory report prepared for the task force noted that a study team "undertook 140 program reviews encompassing 218 separate federal and federal/provincial programs." The report's conclusion was that Canada had created a system of "giving with both hands," where new programs were constantly created without ensuring that similar programs didn't already exist in other departments or provinces, and with very little understanding or review of tax measures by subsidy program officials. The upshot of this ignorance was that "very few people can identify the extent of the programs' operations, or their impact."

The task force Identified six pervasive issues across government. The first was "universal subsidy," the problem of using subsidies as the first response to any problem, and their accumulation in several forms: tax breaks, permanent statutory programs, and discretionary line items in annual budgets. This accumulation created the second problem of "fiscal totality." At the time of the task force's review, only 21 percent of subsidies (across government, not just to business) were "non-statutory" items. This meant that from year to year, just shy of 80 percent of federal subsidy programs proceeded essentially without oversight. That programs usually crudely tracked inputs and outputs instead of tracking outcomes did not help matters.

The third problem identified by Nielsen's task force was the uselessness of program review. Programs were created and departments duly reviewed them on a routine basis without asking fundamental questions as to why a particular program existed—they merely looked at how effectively and efficiently a government lever operated, instead of looking to see if the mechanism itself was worthwhile.

These first three problems created permanent momentum for subsidy programs. The only thing in government easier than setting up a new program to write subsidy cheques was to keep an existing program going. Three other issues compounded the impact of this already-deadly trifecta. A lack of institutional memory within government prevented the emergence and development of deep knowledge and expertise. The measurement of inputs and outputs as opposed to actual outcomes disincentivized cost control. And a relentless bias to the status quo worked hard against change.

Departments and agencies set up to administer government business support programs did not know how to assess business cases. They were not helped in this regard by vague eligibility criteria and routine overstatement of project benefits by applicants. The auditor general highlighted a case in 1985 of a company with two grant proposals under consideration that included the *combined* economic benefits of the programs in *each* application, as well as another where a proposal for marketing a conference claimed that it would create 450 jobs and generate $450 million in sales.

The services and subsidies study team found that when it came to tax breaks for business, "only [Energy, Mines and Resources] and certain quarters of DRIE can be described as at all tax-sophisticated" and that the rest of government "operates as if there were no corporate tax system" when it came to the design and operation of programs. This meant that tax breaks faced significantly less oversight than spending programs. They were essentially ignored for years, despite making up a tremendous share of overall business support. Throw in naked political and partisan preference, and the result was essentially anarchy.

One DRIE bureaucrat observed in the 1980s that empirical evidence to inform policies was in short supply: "What's fascinating is that the policy process is so disconnected from fact. Very often policies, ideas, programs, and so on are dreamt up in somebody's head, and they're based on a very impressionistic, personalized, fragmented view of industry which is gained just through a process of osmosis. . . . But you know the notion of somebody sitting down, and analyzing the industry, identifying the principal issues and problems, ranking them, analyzing the priority ones in depth, and so on, just is not the case. It doesn't happen. The time frames do not allow it. The information is so uniformly bad, it just doesn't happen."

It was clear, after decades of activist economic intervention, that Ottawa was not very good at it.

How did a government staffed with intelligent, talented public servants keen to rationalize its approach to the economy end up in this situation?

In the late 1980s, an anonymous bureaucrat at DRIE vented to public administration academics Michael Atkinson and William Coleman that "for the past ten years, every time government has had a problem, rather than trying to address the problem, they've created new organizations to deal with it. And you've got organizations overlapping organizations, overlapping organizations. It just simply means that responsibility is so diffused that nobody is responsible. I defy you to find one person in government who is responsible for one thing. It just doesn't exist anymore. It's just a chaotic mess."

A Finance official fumed, "If I were an outsider looking at the federal government, I would be totally disgusted with what is going on inside. It's ridiculous . . . Every government department has people that have their little fingers in the policy end . . . You have all these different agencies floating off in this big netherland, going off in different directions without any sort of co-ordination."

The reality that Atkinson and Coleman discovered In *The State, Business, and Industrial Change in Canada*, their seminal work on Canadian economic policy and its intersection with public administration, was that the government was ill-equipped to deal with the challenges of formulating a nimble, proactive industrial policy. It had too limited a toolbox. Anticipatory or proactive industrial policy requires a strong state with a public service empowered to act independently in economic matters, as well as strong and centralized national business and labour associations that can bring their own expertise to the table and effectively represent those sectors to government.

Of course, Canada had a professional and non-partisan bureaucracy over this period. Our public service has historically been well-regarded by bureaucracy connoisseurs the world over. But we also had a strong Westminster tradition of direct political accountability for each decision of government, by which politicians could over-rule professional and non-partisan officials. And they did.

Canadians have a not-inaccurate image of the business sector as well-represented in Ottawa by lobbyists and executives. Business has always been able to play a large role in the internal politics of the two parties of government, aided through most of our history by direct political donations. Most of this, however, has been driven by individual firms or sectors on an ad hoc basis, as opposed to organized and sophisticated business associations. That has limited the degree to which the broader economy, as opposed to individual firms, have benefitted.

Our labour movement is weak relative to, say, Nordic social democracies, where they play a direct role in economic adjustment policies. The lack of labour pressure, or a credible socialist threat, has effectively removed working people from the economic conversations that wind up delivering corporate welfare. In fact, labour unions, focused on the job security of their members rather than broader economic security, have historically been accommodating of business subsidies.

It doesn't help that Canada, like the United States, but unlike most other advanced democracies, has never developed a political and party system polarized around a self-conscious industrial working class on the one hand, and capital on the other. Intertwined sectional, language, and religious concerns have always predominated.

The inability of government to bring business and labour or working-class representation into the formation of industrial strategies in the post-war decades assured their failure. "When the political conditions for anticipatory policy are not met," Atkinson and Coleman noted, "and anticipatory policies are attempted anyway, frustration will follow. Not only will these policies fail, but the exceptional resources committed will fall into the hands of the strongest" interests.

What you get is not economic development but corporate welfare. The process becomes deeply politicized, with government focused on job-protecting bailouts and ongoing support for weaker sectors—those reliant on government in the form of approvals or contracts, as well as slow-growing geographic regions. The whole mess becomes vulnerable to the lobbying of connected firms or industries, which walk away with the money.

* * *

Another dynamic undermining government attempts to bolster the economy were its periodic reformulations and reshuffles of programs and entire departments, such as we saw in the last chapter. They were the product of interest blocs within the public service itself, along with frustrated political leadership. The Trudeau government intended to bring a new rationality to government when it was first elected in 1968, prioritizing thoughtful policy analysis informed by evidence. While this change in approach succeeded in giving civil servants sexy new policy career tracks at the expense of traditional paths centred on program management expertise, it ran headlong into cultural barriers imposed by three institutions at the heart of government: the Department of Finance, the Treasury Board Secretariat, and the Privy Council Office.

The Department of Finance is in many ways the nerve centre of government, holding the pen for large chunks of each annual budget. It was temperamentally disinclined to wade into industrial policy. When elected officials wanted to spend more on economic objectives, the flinty department faced a choice: it could double down on its role as guardian of the federal budget and fight a defensive war against other departments and ministers over each proposed line item, or it could explore ways to lead on the government's overall approach to the economy.

The department ultimately chose the former. "Departments are constantly coming up here with ill-conceived ideas which would either screw up the economy and/or employ an economic instrument, like taxation, for a social or cultural goal," one Finance official chortled to Atkinson and Coleman in the mid-1980s. "I find it satisfying and exciting to see these policy proposals shot down by our boys purely on the grounds of economics."

The vacuum left by Finance was filled by adventurous officials from Industry, Trade and Commerce, DREE, and other departments and agencies. They staffed up with the type of economists that would once have gone straight to Finance and assumed a significant role in the evolution of economic policy.

The Treasury Board is a committee of ministers that scrutinizes and approves spending initiatives proposed by Cabinet. The Treasury Board Secretariat (TBS), which was hived off from the Department of Finance in the mid-1960s, supports the Board and its president and is one of the central agencies of government.[12] In the early 1970s, the Secretariat tried to rationalize program evaluation and Cabinet approval processes using trendy cybernetics research that betrayed a real naiveté about political decision-making. The planning branch of the Secretariat failed to win over even the programs branch within the Secretariat, let alone the Department of Finance or the Privy Council Office, which was also jealously protective of its prerogative of informing Cabinet deliberation.

The Privy Council Office[13] (PCO) under Clerk of the Privy Council Michael Pitfield[14] developed its own forward-looking and process-oriented system for economic planning. Much looser and vaguer than other approaches, it attempted to provide structure for Cabinet to consider big issues and enable the politicized horse-trading that characterizes actual policy making around a Cabinet table.

The upshot was that the three key players in economic planning were pursuing mutually incompatible cultures of planning, creating a morass into which industrial strategists would sink for years to come. As Gordon Osbaldeston, the later deputy minister of Industry, said, "I wasn't really anxious to go into the marsh into which a number of my predecessors dashed to disappear forever."

While Finance, the Treasury Board Secretariat, and the Privy Council Office were talking past each other, ministers of other departments more or less disregarded Finance's fiscal frameworks and went around Treasury Board to announce programs or initiatives once they had been approved

---

12 A central agency, in Ottawa parlance, along with Finance itself, and the Privy Council Office. The Prime Minister's Office is often considered a central agency as well, despite not being formally part of the professional public service.

13 The Privy Council Office is the department that nominally exists to support Cabinet and its deliberations. It in practice also acts to some extent as the 'Department of the Prime Minister' and helps quarterback key initiatives within the public service.

14 Pitfield became a key Trudeau ally in the bureaucracy and was promoted to clerk of the Privy Council, the head of the public service, in 1975.

in principle by Cabinet. This led to the establishment of programs or initiatives with only rudimentary departmental analysis and unreliable costing to guide their implementation.

Many economic programs also lacked clear and defined goals. Planning should ideally begin from a presumption or goal with regard to *ends*. The arena of democratic and bureaucratic politics is ultimately an arena of *means*. At the practical level, there is no way to disentangle means and ends without creating political winners and losers. In a policy environment where politicians are hesitant to be explicit about inevitable tradeoffs, and bureaucrats retreat to safety, the inevitable outcome is programs launched with vague hopes of discovering their goals during implementation.

* * *

Efforts to formulate a high-level industrial policy came and went through the difficult environment of the mid- and late-1970s. The imposition of wage and price controls, the election of the Parti Québécois in Quebec in 1976, higher inflation and interest rates, and budget cuts from 1978 onwards left many Liberals focused on everything but optimal policy design. Regardless, regional development continued to be a priority and brought its own set of problems.

The Trudeau government folded its Department of Regional Economic Expansion into Industry Trade & Commerce to create the Department of Regional Industrial Expansion (DRIE). The new department oversaw the Industrial and Regional Development mega-program (IRDP). A huge program with vaguely defined industrial and regional development goals, IRDP could be used for research and development, capital expansion, marketing support, and every other conceivable business objective. To facilitate the regional development objectives of the program, each census division in the country was assigned to one of four development tiers, with more deprived regions eligible for more support.

Despite this intricate architecture, analysis of the program found that the program actually largely ignored the tier system, while bureaucrats took advantage of the lack of explicit commitments in the program to distribute capital in ways that had nothing to do with its original design. Nielsen's study team pointed out the fundamental incoherence of IRDP, writing that "[it] attempts to address both industrial development objectives and regional development objectives. Unfortunately, national economic efficiency does not easily fit with interregional equity and the conflict in objectives has bedevilled the program from the start." Broad objectives such as those of the IRDP "prohibit nothing, and in consequence, almost any proposal can be said to advance one or more of them."

The inability of programs to cohere with themselves, much less with each other, bedevilled any attempt to bring them together into a coherent macro-policy. This is ultimately a political failure. The public service contributed, but responsibility rests first and foremost with elected officials declining to be honest with each other and with voters about priorities and trade-offs.

* * *

The complexity of the constantly shifting landscape of subsidies to business consistently undermined its own effectiveness and made measuring the impact of individual measures all but impossible. "Right now," the Nielsen study team found, "the system is a veritable gold mine for entrepreneurial 'stackers' of benefits, and a discouraging morass for other firms who don't even apply for assistance, let alone stack." The Macdonald Commission on the Canadian economy, launched by Trudeau in 1982 and reporting to Mulroney in 1984, reached a similar conclusion, arguing that "the contribution of corporate tax incentives to investment decisions is difficult to assess because these incentives often change substantially over time. Indeed, some observers argue that the uncertainty generated by frequent modifications to the corporate tax law significantly discourages investment."

An Economic Council of Canada paper from 1980 on the effectiveness of subsidy programs concurred that rapidly changing programs and agencies made them less effective. What subsidizers in government rarely understood was that to actually incentivize investment, subsidies had to be clear, predictable, and simple, and represent a substantial fraction of the cost of the investment. Complexity appeals only to large companies able to spend the administrative energy to stack benefits, but these are the firms least likely to have their actual investment decisions determined by the addition of government funding. The other class of companies eager to do the legwork to obtain public money are those that are unprofitable and at risk of failure.

Another important finding of the Economic Council of Canada authors was that the rationalized culture of bureaucratic planning clashed with the animal spirits of the market. Government planners, they argued, "largely ignore the non-financial causes and barriers to new investment," and that "there are several important causes or barriers which might be called personal, emotional or even sometimes irrational" that are actually the main drivers of business investment decisions. Program designers were making "mechanical assumptions" that subsidy dollars would lead to businesses to invest, not realizing that business is more alchemy than chemistry.

Given all these failings and obstacles, it was inevitable that the efforts of successive post-war governments of different partisan stripes foundered in their efforts to engineer prosperity. It was also inevitable that persistent failure would lead the government toward a brutal fiscal reckoning. That came in the 1990s, as Jean Chretien's government cut Canada's social welfare state to the bone to appease investors in Canadian public debt after decades of profligacy. To take a partial inventory, regional development agencies suffered hundreds of millions of dollars in cuts from 1992 onwards, the government eliminated several preferential tax credits, and research and development subsidies went from over $700 million in 1994-1995 to under $250 million by 1997-1998.

The long shadow of the 1990s cuts reached far beyond corporate welfare to missing units of public housing, the deterioration of health care, and the downloading and elimination of social services. The corporate welfare state could not be kept down for long. A new golden age was just around the corner.

CHAPTER SIX

A CLEANER, SMARTER TROUGH

In 2014-2015, Canadians were forking out roughly $14 billion in federal business subsidies. The number for 2023-2024 is expected to reach, on an apples-to-apples basis, $33 billion, and as much as $40 billion with climate-related subsidies factored in—more than at the peak of the Mulroney era.

The recent history of climate-related subsidies and innovation policy is instructive. There was a time when government subsidies to business intended to address climate change were largely directed at reducing emissions. They were nevertheless controversial: in 2019, $12 million from the ($2 billion) Low Carbon Economy Fund went to profitable grocery chain Loblaw to buy higher-efficiency refrigerators and freezers. Certainly, if you are taking the perspective of the emissions-reduction planner, helping one of Canada's largest grocery chains reduce emissions across its national footprint makes sense. If you're a small grocer and a competitor to Loblaw, I would imagine that seeing the federal government helping your large, profitable rival eat your margins with your tax dollars smarts a bit. Programs like this are also criticized, with some justice, by environmentalists as subsidies to fossil fuel companies to continue and expand production. Accounting for climate-related business subsidies in the overall envelope of economic subsidies at one point led to ambiguous judgment calls.

There is much less ambiguity now. We mostly left them out of the figures at the beginning of this book, but the government has explicitly moved from a focus on decarbonizing existing industries to industrial strategy investments in emerging sectors.

The government created an $8 billion Net Zero Accelerator fund, focused mostly on decarbonization and transformation of industry, although it also includes clean tech funding, especially focused on support for the development of electric batteries. In 2022, the Liberals announced the $15 billion Canada Growth Fund, which was retooled after the passage of the US *Inflation Reduction Act's* green investment subsidies to provide a floor for Canada to compete with American largesse. The same budget saw $3.8 billion allocated to the Critical Minerals Strategy. Budget 2023 included $11 billion in new or enhanced green industry tax credits, and Budget 2024 added another $7.2 billion in clean electricity tax credits.

All of this spending is spaced out over multiple years, but we are a looking at a significant new ecosystem of business subsidies with their purpose shifting, to varying degrees, from a laser focus on reducing emissions to industrial policy. They have also grown enormously. Clean economy subsidies have gone from 1.6 percent of the total to 18.3 percent since 2015. That's not necessarily illegitimate or wrong, and it's certainly gratifying to see the government take the climate emergency seriously. But as we've seen in previous chapters, these things are hard to do well, and the magnitude of the new programs and the speed with which they are rolling out should concern us. That they're given political cover by the need to respond urgently to climate change does not eliminate the requirement to get this right.

That takes us back to the single biggest package of federal government business subsidies in our history: the announcement in spring 2023 of subsidies of up to $28 billion granted to European auto giants Volkswagen and Stellantis to build electric car battery factories in Southern Ontario. The package matches the American *Inflation Reduction Act* (IRA) subsidies and includes $13 billion for Volkswagen and $15 billion for Stellantis (with one third of the latter covered by the Ontario government) in production tax credits over the course of a decade. Volkswagen also received $700 million from the federal Strategic Investment Fund and $500 million from the provincial government in up-front capital to offset the cost of its

$7 billion investment. The Volkswagen subsidies alone, the prime minister told Canadians, would create around 3,000 direct jobs and up to ten times that number in indirect jobs and generate $200 billion in value over its lifetime.

If you're an optimist, you might see the market for electric vehicles poised to grow dramatically over the coming decades. Canada has a mature auto sector deeply integrated with its American counterpart. Auto sector investments tend to be long-term and often result in follow-on investments from companies looking to co-locate facilities, particularly in this case, where a major foreign automaker is coming to Canada for the first time. Auto supply chains attract other investment by third-party suppliers. Mining companies can be sure of steady demand for products like lithium and manganese, giving them headroom to pursue their own cost-reducing or value-adding innovations. There could be substantial investment in research and development around batteries and critical minerals.

You might also be pleased that the risk in this deal is back loaded, meaning that most of the funds won't be committed until the factory is up and running and producing saleable batteries. You'll tell yourself that the US IRA and its hundreds of billions of dollars in offered subsidies make this the minimum buy-in to be a competitive location for this kind of investment. And who's to say you're wrong? It's certainly possible that this could turn out to be a great investment that pays itself back in five years, as Minister Champagne told Canadians to expect.[15]

If you're more realistic in your outlook, you will note that is a huge bet on two facilities owned by two foreign firms and, more precisely, that their presence in Canada will have a transformative downstream effect on the entire auto and minerals value chain in this country.

The history of these kinds of bets is that when they don't fail outright, their projected economic multipliers turn out to have been dramatically overstated and the public money invested ends up subsidizing labour costs for companies without adding much in the way of broader value.

---

15 I will be charitable and assume he meant after five years of operation, since the factory is only slated to come online in 2027.

In the wake of the announcement, economists pointed out that the government's assumptions about the impact of the subsidies used fairly simple input-output modelling as opposed to more sophisticated modes of economic modelling, as well as an implausibly high indirect job multiplier. Past multiplier figures claimed by the government, like the anticipated return of $25 for each dollar invested into the Innovation Superclusters Initiative, have proven (at least so far) to be complete fictions.

One auto sector expert, Greig Mordue at McMaster University, expressed skepticism that the Volkswagen and Stellantis investments will provide significant value to the critical minerals miners and processors, given the lag between turning proven reserves into operating mines. Even in the government's rosiest scenario, Canadians are paying $400,000 per job over a decade. Unless there are monstrous positive spillover effects from the investment, that won't be a very economical use of taxpayer money. This is especially the case in a tight labour market where jobs are not particularly scarce to begin with. Remember, local employers will have to compete with VW and Stellantis, backed by all that government subsidy. There is also some risk that lithium-ion batteries are supplanted by other technologies in the coming years.

That Volkswagen and Stellantis are foreign-owned companies is an important factor. Profits from the sale of batteries produced in St. Thomas and Windsor will mostly go to foreign shareholders. They won't be staying in Canada. That means the shareholders who get rich on these investments, presuming they succeed, will be paying taxes, endowing university research chairs, museums, or wings of hospitals or investing in other countries, not ours. Shuffling around Canadian public money to pay for a new coat of paint on the Rijksmuseum in Amsterdam (where Stellantis is headquartered) seems suboptimal.

The question of whether this investment is justified has several layers. Net benefits are one measure. Will this massive subsidy, once all downstream effects have been calculated, provide more to Canadians than was spent? Another measure is opportunity cost. What else could $28 billion over ten years have accomplished if spent differently? Instead of entering a competitive subsidy race with the United States, might we have invested more in other areas of current or emerging comparative advantage, or health care, housing, or a number of other worthy causes? Or just saved the fiscal firepower, to use a term the government is fond of, for another day?

These are not easy questions to answer, and it is genuinely discouraging that the best we have got from the government on these questions is mere assertion: 3,000 direct jobs! $200 billion in value! Confident countries, as the prime minister has said many times over the years, invest in themselves and in the future. Confident countries, it seems to me, should also not be afraid to show their work when betting eleven figures' worth of public money on a megaproject, no matter how the deal is structured.

Opposition parties were also worryingly quiet about this. The politics of job creation are irresistible to vote-seekers of all stripes. Conservative leader Pierre Poilievre meekly asked the Parliamentary Budget Officer to do some homework on his behalf. The NDP's Jagmeet Singh, after months of attacking the Loblaw Corporation and its unsympathetic president, Galen Weston, for pandemic profiteering, said nothing. The most the NDP, the party that coined the term "corporate welfare bums," could be stirred to do was thank the government for "finally step[ping] up and start[ing] working toward securing Canada's climate and economic future," before making a perfunctory bleat about ensuring the plants are unionized. And that was for just the Volkswagen plant. The opposition parties, when the Stellantis deal details were made public in early July 2023, opted to let it pass without so much as a press release.

The historic weakness of our governments at rationally executing on even simple economic development policies should make us skeptical of the Volkswagen and Stellantis deals, especially when all this policy's biggest boosters have to offer is cheerleading of the sort that encouraged Prime Minister Laurier to subsidize two new transcontinental railways that went bust in about a decade. These electric vehicle investments have been followed by billions more for battery maker Northvolt and Japanese automaker Honda. In more challenging economic circumstances for the EV industry, many of these investments were scaled back or put on hold over the course of 2024. Sweden-based Northvolt went into bankruptcy protection by the end of the year, leaving the fate of its Canadian investment an open question. Days before the end of December, Minister Champagne acknowledged that the era of the mega-deal was at an end and that policy would have to move to "consolidation," including shoring up upstream weaknesses in our natural resources supply chain.

It is not too much to ask for our governments to be clear about its assumptions and estimates. If they don't exist or are written on the back of a napkin, politicians shouldn't be committing enormous sums to these projects.

* * *

Another major theme of government economic development energies has been high-tech sectors and niches within sectors, including advanced manufacturing and digital technologies such as AI and quantum computing.

Jaded observers might recall the bizarre spectacle of Prime Minister Trudeau glibly explaining the basics of quantum computing to amazed reporters in front of a *Good Will Hunting*-esque blackboard in 2016 and dismiss the government's focus on emerging technologies as either a cynical exercise in brand-building, or a simple fascination with novelty, or a sinister desire to undermine natural resources industries.

There's something to the cynical brand-building and naiveté, but I believe the government's support of high-tech fields is based on genuine enthusiasm for Canada's potential to be a player in these fields. There are good reasons to think it can be. Technology fields depend on human capital. Canada's workforce is among the most educated in the OECD, even if the 'most educated' statistic often bandied around is inclusive of college diplomas as well as advanced degrees and may overrepresent the supply of engineering and other talent in the country. We punch above our weight in global research leadership with strong universities that attract smart students from all over the world. We've benefited over the life of the Trudeau government from Brexit and the Trump administration, which made Canada a more attractive destination for researchers and students looking for rich-world Anglophone university at which to work or study, at least in the short term.

Research-intensive tech fields are also a place even the stodgiest conservatives can concede benefit from government intervention. As a rule, a free market left to its own devices will under-invest in research and development. Conventional economic theory teaches that a company can't capture the full return on its research. And so Canada, as we have seen, has for generations gone beyond training research talent at universities and colleges to directly subsidize research and development through measures like the Scientific Research and Experimental Development (SR&ED) tax credit, the Industrial Research Assistance Program (IRAP), National Research Council (NRC) labs and facilities, and various tailored research and development support programs at ISED and other departments.

Yet we've seen nothing of the expected world-leading industrial competitiveness and productivity gains supposedly awaiting us on the technological frontier. The reasons for this have been the subject of considerable argument for many decades. The Science Council and the Economic Council of Canada, advisory bodies to the federal government active from the mid 1960s to the early 1990s, represent the poles of this debate. For much of its history, the Economic Council claimed tariffs and other protectionist policies were barriers to innovation and productivity growth—barriers that cultivated an unambitious managerial class slow to make use of the latest technologies. It called for trade liberalization as a response. The Science Council found the roots of Canada's productivity problem in a stunted industrial sector dominated by foreign companies that kept Canada in a quasi-colonial, dependent economic relationship with the United States. It called for a "technological sovereignty" approach to break the cycle of dependence. Canada, as it often does, chose a middle way: substantial trade liberalization along with continued public support for research and development that would hopefully lead to productivity-enhancing innovation.

Today, Canada's financial support for private research and development is quite generous by global standards. As of 2020, we spent just over a quarter point of our GDP on it, with the lion's share coming from federal tax incentives. We are more generous than the OECD average, not to mention such innovation heavy-hitters as the United States, Sweden,

Israel, Finland, and Germany. We are about on par with the Netherlands. Out of all those economies, however, ours lags behind in both GDP per capita and GDP per hour worked. We also work more hours per person than all the aforementioned countries except the United States. We are spending more to get less for our money and our time, and businesses themselves invest less in industrial R&D than in the other countries.

This is not a new reality. Since the 1990s, major budget speeches from finance ministers Don Mazankowski, Paul Martin, and Chrystia Freeland have all signalled commitments to address lagging business innovation and productivity. And the Justin Trudeau government has spent heavily on making Canada more innovative with new measures like the $950 million Innovation Superclusters Initiative and the $4 billion Canada Digital Adoption Program (CDAP). There is also the flagship $7 billion Strategic Innovation Fund (SIF). Like their predecessors, unfortunately, all these programs suffer from a lack of clear priorities and all have struggled to make a measurable difference.

The Strategic Innovation Fund was created by Industry Minister Navdeep Bains in 2017 by merging existing large-scale business support programs for the auto and aerospace sectors with dedicated, large-scale research-and-development grants. The program is now the government's flagship one-off business support program. It has spent around $7.6 billion since 2017 on various projects. By the time you read this, the number is likely to be higher. The program has been criticized for just how discretionary it is: there is little indication that it follows any kind of set investment strategy. Rather, it responds to political priorities. Batteries and critical minerals weigh heavily in last year's announcements. Biomanufacturing and life sciences dominated the COVID years. The NAFTA renegotiation years saw considerable sums poured into the steel industry.

The SIF has also been criticized for excessive focus on supporting foreign firms instead of domestic ones. Analysis of SIF funding through early 2019 bt *The Logic*, an innovation economy news publication, found that a hair over half went to foreign firms. In the intervening years, that figure has crawled up to just over 60 percent of listed projects (I specify listed projects because a few investments that have not panned out, such as contributions to the Canada Kuwait Petrochemical Corporation, Element AI, and smart-glasses-maker North, no longer appear on the department's list). While there is nothing wrong with foreign direct investment as such, there are reasons to be concerned about subsidizing the research and development efforts of foreign companies if the value of their products or services is based on intellectual property. The positive spillovers of a large factory are somewhat obvious in the community in which it exists. The same can't be said for positive spillovers of corporate research that generating patents and other IP for a foreign-owned firm. In an economy dominated by intangible assets, the spillovers might even be negative if the profits from that IP are realized abroad and the foreign companies constrict the freedom of Canadian firms to operate.

It is hard to avoid the conclusion that the purpose of SIF is merely to signal the government's concern for specific sectors of the economy or regions of the country. It has little to do with lasting economic transformation. As we've seen, a lack of focus has bedevilled large, politically sensitive subsidy programs for generations, leaving the officials in charge of administering them in a weak position to push back on their political masters with political objectives.

Budget 2017 also allocated just under $1 billion for what the government called the Innovation Supercluster Initiative, an attempt to leverage well-studied cluster dynamics to make Canada's economy more innovative. Clusters, innovation policy scholars argue, are dense interconnections of educational institutions and companies of all sizes; they exist in specific places and create value from the active ferment of people, ideas, and money. Silicon Valley is the archetypal cluster. Authors have been writing books about how to create the next Silicon Valley for decades. The original Silicon Valley benefited from billions of dollars in government (especially defence) contracts to build up the modern capital of global high-tech industries. A $900 million investment by Ottawa is not really of the same order of magnitude as what the US invested in Silicon Valley. More baffling still is Ottawa's decision to split its investment among five so-called "superclusters" evenly distributed among Canada's regions and each targeting a different sector of the economy. Whatever one may think of Silicon Valley, we are all at least aware that there are not five of them.

The Trudeau government claimed early on that this program would ultimately add $50 billion to Canada's economy—a multiplier of twenty-five once private sponsorship of the clusters was factored in. We were also promised 50,000 new jobs. A Parliamentary Budget Office report from 2020 found that the program would create roughly half that many jobs and new economic activity of about $18 billion (and it was still using generous multiplier assumptions).

Many ISED officials believed the Parliamentary Budget Office report was an unfair snapshot taken at an awkwardly early point in the program's development. Former ISED deputy minister John Knubley wrote a report for the Brookfield Institute in 2021 defending the program. He pointed out the very real problems with Canada's innovation policy prior to the creation of the superclusters and argued that "just doing the same old things would be very costly in terms of innovation and productivity performance . . . [the superclusters] were seen as a fresh type of innovation policy, with objectives directly linked to traditional weaknesses of Canada's innovation performance." Both reports are interesting reading, but the central problem of the superclusters initiative is simply that to create dense and productive linkages between business and post-secondary institutions that would help substantially correct Canada's historic weakness in innovation, the government proposed to split a relatively small sum of new research money five different ways.

That the government was never convinced of its multiplier math is indicated by the relative paucity of its investment. If officials genuinely believed they could generate a return on investment of $25 for each federal dollar it spent, why stop short of $1 billion?

Ottawa appears to have lost faith in the clustering dynamic that originally inspired SIF, as well. With a cash top-up and rebrand as "Global Innovation Clusters" in Budget 2022, the program de-emphasized the need to build interconnections in specific places. The Nova Scotia-based Ocean Cluster is now funding research projects in Kitchener-Waterloo and Calgary. These initially place-centred bodies are now essentially small-scale sectoral research granting councils.[16]

There are other examples. Much could be said, for instance, about the consolidation in the 2018 budget of regional development agency programs into one department, followed by their decentralization and the creation of three new agencies[17] in the 2022 budget. But let's focus instead on the Canada Digital Adoption Program (CDAP). It doesn't rank among the government's sexier, research-intensive policies, but it represents a huge financial outlay—more than four times the size of SIF.

The government announced this $4 billion program in Budget 2021 to help Canadian businesses bolster their web presence and digital offerings, such as online storefronts. While it is true that in decades past, Canadian business has been slow to adopt technology, the announcement came a year after the beginning of the global pandemic, and it wasn't until a full year later, in 2022, that the program's structure and terms were made clear. Businesses had already spent two years moving online in record numbers.

The program offered small businesses two funding streams. First, there was the Grow Your Business Online stream. Businesses could claim a small grant of $2,400 from recognized provincial providers to develop an e-commerce strategy. For some reason, these providers were also expected to create youth placements for business advisory services that would be subsidized by the program.

Then there was the Boost Your Business Technology stream. Businesses could get grants of up to $15,000 to develop a digital adoption plan with advisory services sourced through a Digital Adviser Marketplace created by the government. Once the business had completed its plan, it could apply for a loan of up to $100,000 from BDC to implement it. As with the first stream, there was a youth angle where businesses are eligible for a wage subsidy to hire a student to help them. I can only assume this angle

---

16 I am indebted to the work of Alex Usher and Paul Wells on this file.
17 Pacific Economic Development Canada (PacifiCan), Prairies Economic Development Canada, and the Federal Economic Development Agency for Northern Ontario (FedNor) was spun out from being a mere ISED branch to a full agency.

played to the prime minister's obsession with youth service programs which, as we saw with the WE Charity scandal during the pandemic, have not always worked out for him.

Like the Scientific Research & Experimental Development tax credit, the Canada Digital Adoption Program was administratively complex. As a result, much of the value accrued to middlemen, particularly the consultants required to apply and comply with the program's terms. Leaving aside that massive inefficiency, it is hard to explain why the program existed. After two pandemic years, any business that had not digitized some of its operations probably did not deserve the help. And beefing up online was not fundamentally difficult or expensive thanks to the plethora of private sector offerings available—not least Canada's own Shopify.

In spring 2023, the *Globe & Mail* found that just $131 million in Canada Digital Adoption Program grants and loans had gone out the door. It was shelved permanently a year later, with just over $500 million of its funding envelope committed. Would it have been better to leave this program as a briefing note on someone's desk? Perhaps. But once the money was announced in the budget, and lobbyists and journalists called for follow-up information, the government seemed to feel that it had a constituency to please by rolling it out.

Between SIF funding that has failed to define coherent objectives beyond good headlines, a superclusters initiative that tried to spread too little funding over too many priorities to make a difference, and a mammoth, overcomplicated small business support program that not many seemed to want, the Trudeau government has struggled mightily to make real inroads on changing Canada's innovation landscape.

The Liberals have managed to more than double the amount of federal subsidies to private business since taking office in 2015. As the government has aged, the sums have got bigger as the justifications have got thinner. Leaving aside its annual budgetary deficits, an issue to which more than enough ink has been devoted, there is no clear evidence that the money is being spent in ways that will make us all richer or that will create new, good jobs for people. In fact, no one in government is keeping track of how many jobs are being created.

Given Canada's long experience of subsidizing business to no lasting economic effect—all the previous examples of short attention spans, political hype, lack of directional planning and institutional expertise, the difficulty of separating means and ends, and the overgrown and confusing patchwork of programs blunting even well-designed and well-administered measures—one might expect a government today could introduce an effective, thoughtful program. The fact is that it is genuinely hard to make even marginal changes to a country's economic fundamentals.

Wedged between the American colossus, industrializing Asian countries, and the European bloc, Canada has to define an economic future for itself. Government will have a role, one way or the other—not even today's Conservative Party believes otherwise. The choice is between doing it well and doing it badly.

The good news is that the experience of other countries demonstrates that smart policy can add up to big change. If it were impossible, South Korea would still be exporting cabbage and fish instead of advanced consumer electronics and pop music. It would not be catching up with the United Kingdom's per capita GDP. It took generations of consistent effort on Korea's part, including focused and patient government involvement, but it is possible.

CHAPTER SEVEN

BLUEPRINTS FOR THE FUTURE

The question of industrial policy, as the Macdonald Commission said in the 1980s, "is not whether a country should have an industrial policy; whether by design or by default, a country will inevitably have an industrial policy of some kind. The relevant question is to what degree industrial policy should favour some sectors over others. Should government endeavour to be, in some overall sense, neutral . . . Or should it attempt to identify and promote the industrial activities in which the country has, or should have, a comparative advantage? . . . If Canadians conclude that a targeted industrial policy is indeed the best choice, does it follow that our government is well placed to devise and implement such a policy?" These are live questions today.

In the wake of the Trump presidency, the COVID-19 pandemic, and the Russian invasion of Ukraine, the Liberal government has changed gears on economic policy. Elected in 2015 as cheerful evangelists of globalization and free trade, their early commitments to investing in the economy were orthodox: new infrastructure and advanced education. As late as 2018, there was serious discussion in Canada about pursuing some kind of trade agreement with China.

Then COVID-19 demonstrated the brittleness of just-in-time supply chains, and the Ukraine war, and increased tensions with China showed the downsides of economic integration with geopolitical rivals, and the Trump presidency's America-first policies demonstrated the potential fickleness of allies. In an October 2022 speech in Washington, DC, Deputy Prime Minister Chrystia Freeland talked at length about the need to "friendshore" critical manufacturing capacity and supply chains to reduce reliance on China and Russia. That wasn't the first time she had said it. And the reliably orthodox Business Council of Canada was in agreement. It shows how quickly and thoroughly the conventional wisdom can change.

The changes renewed government interest in industrial policy after decades of relative neglect. For much of the last three decades, conventional wisdom had been that there was no need to tinker with subsidies. An unleashed free market and even more liberalized trade would solve our economic problems. That approach has given way to a recognition that to some extent every developed economy in the world has had its industrial structure determined, for better or worse, by government.

Two of the better examples are Taiwan and South Korea, countries that entered the post-war world poorer than the global average, yet have since developed prosperous, liberalized market economies with the state playing a crucial role. It is hard to imagine Taiwan enjoying its considerable prosperity and its continued American security guarantee if it had not managed, at great public expense, to specialize in semiconductor manufacturing, placing itself at the centre of the global value chain of the digital economy. No one looking at Taiwan in the 1960s would have spotted any latent potential for high-tech manufacturing, and there was nothing magical about Taiwan's people that made them peculiarly well-suited to capture this space. It was a direct result of deliberate government policy.

Finland and Sweden similarly advanced from reliance on agriculture, forestry, and resources to highly productive economies dominating in advanced fields, home to some of the giant global corporations that the Canadian government is keen to entice with subsidies (and they still have strong resource sectors). Once you accept that a given country's suite of advantages can be intentionally changed, it becomes asinine to deny that government *can* have a productive role to play. Of course, it doesn't always work. Malaysia has tried to mimic what it saw work in Taiwan and Korea and did not do as well.

Money is a factor, but not a deciding one. Other nations, including us, have spent more to accomplish less. The important thing is that successful countries mostly spent their money well while properly aligning their politics and institutions. As the Macdonald Commission pithily put it, "resource misallocation is an inevitable consequence of government support programs, just as resource misallocation is an inevitable consequence of private sector decisions. The challenge for Canada is to do at least as well as its major competitors in the exercise of industrial policy."

It is important to remember that governments pursue industrial policies because we live in a global economic system in which capital is extremely mobile. The free market in capital will often over-invest in some sectors, and under-invest in others. Governments use an array of carrots and sticks to entice and cajole the owners of capital to use it in different ways than would otherwise be their preference. So industrial policy is not pursued for its own sake: it is an attempt by government to compromise with the power of capital to go elsewhere. A lasting compromise, with government working in concert with industry, can help create lasting wealth and a better life for citizens.

So what should Canada do? One of the reports undertaken for the Macdonald Commission offers a strikingly lucid analysis of the options available to small trading economies in the era of the knowledge economy. Economist Richard Harris' study, subsequently published as *Trade, Industrial Policy and International Competition,* remains one of the most forward-looking contributions to Canadian public policy.

Harris showed that the choice between trade liberalization and industrial policy was a false one. Freer trade, on the whole, was good for Canada, but Canada's small domestic market meant that government policies had to focus on the rapid scaling up of businesses to export internationally. Policy, argued Harris, should focus on industries with low barriers to entry and with high returns on human capital—typically those in high-tech, high-productivity fields. It should avoid industries with large barriers to entry, where subsidy competition between countries creates a transfer "primarily from the taxpayers of the winning country to the shareholder of the multinational." (Sound familiar?) Getting the most from globalization should be a two-step process: opening markets, and setting the policy and institutional mix to enable your country's success in world market niches that will bring back outsized returns.

What makes innovation-focused, R&D-intensive high-tech sectors so special? In Harris' view, policymakers undervalued the connection between innovation and the winner-take-most market structures first proposed by Austrian economist Joseph Schumpeter.[18] Firms and their home countries see economic returns to innovation when they are the first big entrants into a new niche and are able to build moats around their business, for example, by owning key patents or benefiting from network effects that enable a tech service. Facebook, Google and Amazon are all good examples of innovation economy winners, if not particularly sympathetic ones, their services having real negative effects on consumers and workers. Finland's Nokia and Sweden's Ericsson are also excellent examples; Canada has been eager to pay both to invest here.

The goal, then, should be to help Canadian firms compete in the global innovation economy, where one or several could break through into strong global positions and scoop up big returns. Those quasi-rents could then be taxed and re-invested by home governments. The policy levers to make this happen could involve support for commercial research and development, procurement policies, and preferential access to capital. Success depends on strong fundamental research and attention paid to skills and talent in the labour market.

None of this is to say that Canada should ignore other parts of the economy. Indeed, an economic strategy focused *only* on high-wage, high-tech sectors is doomed to produce more inequality and to accentuate regional disparities. This was the Achilles' heel of New Labour's economic policies in the UK. But the industries in and around high tech, including advanced manufacturing, are places where the most outsized success can be found with the marginal federal dollar.

---

18 Schumpeter, famous for his theory of creative destruction, was an aristocratic conservative, but has had a curious afterlife as a muse not only to hardcore libertarians but management and innovation theorists all over the political spectrum.

Winning firms, of course, are themselves at risk of being disrupted. BlackBerry was a homegrown innovation winner undermined by the arrival of the iPhone. This requires governments to maintain strong research and talent bases and always be looking for the next breakthrough technologies, and the next opportunities to help companies that will define emerging global industries.

Artificial intelligence is today's breakthrough technology. Ottawa recognized the opportunity and attempted to leverage Canada's domestic excellence in AI research—pioneers such as Geoffrey Hinton and Yoshua Bengio located here early in their careers—to attract still more talent. When AI began to attract huge investment, however, American companies began offering American salaries to our top people, eroding our AI expertise. Ottawa responded with a funhouse mirror version of the National Policy: it encouraged foreign companies to set up research branch plants in Canada; they siphoned the resulting value, in the form of intellectual property, to their home offices. Of the nearly 250 patents resulting from research funded through Canada's AI strategy, 75 percent are owned by foreign tech giants. Those patents now can be used by foreign companies to keep Canadian companies out of the fields and niches they cover. Canada essentially paid to make our own AI companies less globally competitive.

AI use cases, such as ChatGPT, have since broken into the mainstream. All the buzz is around the AI subsidiaries of Microsoft and Google, despite the important contributions of Canadian researchers to the field.[19] Canadians are not much better off for the hundreds of millions spent.

The government appeared to be taking a smarter approach with its new Canada Innovation Corporation. It is designed to operate at arm's length from government, and to incorporate the genuine technical and commercial expertise of the National Research Council's broadly successful Industrial Research Assistance Program. It has the focused

---

19 To the government's credit, Budget 2021 made changes to the program to include a greater focus on commercialization.

objective of helping companies turn their ideas into products and services. The combination of operational independence, actual subject matter expertise, and orientation toward helping companies generate and commercialize the valuable intangible assets that enable success in innovation markets is genuinely promising in ways that past efforts have not been. Alas, the government recently delayed its launch by a few years.

What's encouraging about the Canada Innovation Corporation, nevertheless, is its recognition that not only the priorities have to be right, but the tools, people, and institutions, too. If there is one thing I learned writing this book, it is that policy is nothing without institutions.

We've seen in previous chapters that subject matter expertise is limited in government, to the detriment of economic programs over many decades. Part of the reason for this is the culture of the federal public service. For the most part, these are hard-working and dedicated people, but they are limited in important ways by institutional constraints.

First, career pressure steers the ambitious toward policy tracks, where they work on files for a few years and then hop diagonally to another department and a new set of issues and stakeholders. Skilled generalists able to rapidly familiarize themselves with new subjects advance quickly in core government departments. Real experts who have spent a decade or more working on specific issues do not.

Second, the public service fishes from a shallow pond. It's not easy to break into government. Hiring processes can be interminable. Qualified mid-career professionals—people with the sort of business and technical backgrounds best suited for economic strategy roles—are not often keen to put themselves through that kind of wait for a job that often does not pay as well the private sector. Bilingualism requirements, to me, are a non-negotiable in this country, but they do serve as barriers to entry for many, particularly when the government's French-language education programs do not seem to produce real fluency. Given these constraints, the government is over-supplied with public policy grads from Ottawa-area universities who started their careers with summer co-ops or internships and never left.

It is not surprising, then, that the government is now outsourcing much of its policy work to the likes of McKinsey. But while outside consultants have a claim to expertise, they are even more secretive than government, more expensive, and totally unaccountable for results.[20]

The NRC's Industrial Research Assistance Program is one of few across government staffed by experts. Its effectiveness has been recognized for decades. Even the flinty Nielsen task force noted simply that "IRAP works." Key to its success is its workforce of Industrial Technology Advisers (ITAs), mostly businesspeople and technical experts hired from outside of government. These people are not that easy to find, and a recent evaluation of the program expressed concern about its ability to replace retiring ITAs. But if the federal government wants to get serious about ambitious industrial policy, it must ensure that career and education tracks exist for policy staff to develop deep technical and business expertise in one or a small handful of areas.

If the government's own policy people are to be successful, moreover, they'll need help. They'll need clear objectives set by elected officials and wide operational latitude to see those objectives done. Economic policy is inevitably political. We can't take the politics out of it, and we don't want to take the politics out of it. The question of what we want our economy to be is inseparable from the question of what we want our society to be. But we can at least reduce how much day-to-day concerns of partisan advantage and electioneering impinge on decision making.

As we saw in the original days of the corporate welfare bums, an absence of institutional involvement on the part of both business, government, and labour results in subsidies flowing to those firms best able to best navigate the worlds of interest group politics and lobbying. It won't be easy to bring these parties to the table. Business in Canada has never felt a need to present a united front. The big firms, especially, are confident they can get what they want from government without any collective action.

---

20 I am indebted once again to Paul Wells for his work on this. See e.g. Paul Wells, "Shine a brighter light on contract government," January 4, 2023.

Only the re-emergence of organized labour as a strong and independent political actor is likely to change that.[21]

None of this will be easy, but if Canada wants to pursue coherent industrial strategies, it needs to make these serious institutional changes inside and outside of government. We have to remember that other countries are managing to make these changes, some of them quite successfully. Even without our level of resource endowments, they have managed to surpass us in the quality of life they offer their citizens. They haven't done it by accident.

21 The flip side of this is that in non-democratic states, business is fine with full employment when the government promises labour repression—this dynamic has been ably theorized and sketched out in Michal Kalecki's classic essay, Political Aspects of Full Employment.

CHAPTER EIGHT

CONCLUSION

This story started in Canada's formative decades with governments flailing amid a series of perceived crises that resulted in the subsidized building of transcontinental railways and the great tariff wall of the National Policy. These mostly served to enrich British financiers and American industrialists, and left Canadians paying through the nose for an absurd surfeit of rail infrastructure controlled by rapacious companies, as well as a rickety industrial base that could not compete in global markets. This gave way to Depression-era experiments in coaxing the economy back to life, then years of CD Howe's program of business-friendly dirigisme during and after the Second World War. The golden age of Canadian corporate welfare followed, from 1950s to the 1980s, a jumble of uncoordinated giveaways whose scale and chaos awed all those who dared look into the vortex. It collapsed on itself in the 1990s, taking a large chunk of Canada's social spending with it.

But the long and inglorious Canadian tradition of corporate welfare refused to die. Successive governments slowly rebuilt it, and the Liberal government of Justin Trudeau did so with gusto, bring us a Green New Deal for shareholders of VW and Stellantis. In eight years, the Liberals have managed to more than double corporate welfare spending from about $14 billion to over $33 billion per year, often repeating the traditional failings of these programs, including unclear goals, an obsession with day-to-day politics, a lack of expertise, and the absence of institutions capable of executing ambitious economic policy.

Today, 59 percent of tax revenues collected by the federal government are paid, in one form or another, as levies on the incomes and everyday consumption of working people. Just 12 percent comes from taxes on corporate income or capital gains. Eroding labour power has contributed to the share of profits in the economy growing from about 12 percent in 2007 to just shy of 20 percent in 2019. We pay a lot while corporations pay a lot less, all the while reaping the rewards of an enormous system of giveaways.

And for all of this, we are not a richer country. We touched on Canada's business dynamism and innovation issues earlier, but Canada's overall economic picture is not rosy. In terms of productivity—GDP per hour worked—we clocked in at just under the OECD average at $53.3 in 2022,[22] marginally behind Italy and just ahead of Turkey and Spain. Our household net adjusted disposable income, a measure of income after taxes and transfers (including health care and other social programs), has us thirteenth in the world after economic heavyweights and tax havens like the US and Luxembourg, but also small countries like Belgium. We look much worse than they do on measures of work-life balance, where we are tied at thirtieth with the United States for leisure time, once again below the OECD average. We are working a lot, relatively unproductively, to earn modest incomes, even when our services like health care and education are factored in.

In a country where housing prices have disappeared into a black hole of unaffordability and where there were 2 million more Canadians without a family doctor in 2022 than in 2019, simply maintaining our standard of living is unlikely if we don't find ways to create more wealth for Canada rather than for shareholders of foreign multinationals.

The era of low interest rates that made bad policy an affordable luxury is over. We have to be far more serious about governing the economy. Further failures will have real costs, and not just to corporate subsidies. Again, the collapse of the first golden age in the budget crisis of the Chretien

---

22 In 2015 USD.

era took a huge amount of our social state along with it. Canadians are certainly feeling the absence of decades of social housing construction now, to mention just one casualty of those budget cuts.

What will happen to childcare if we find ourselves in a new era of austerity? A universal program of affordable access to childcare has been a priority for Canada's progressive movement for generations. The federal government put forward $30 billion over five years to stand up a national system guaranteeing spaces for $10 per day, with nearly $10 billion promised for years following. The program is facing real issues. Daycares are having to pull out of the subsidized program because federal and provincial funding doesn't cover their costs, especially after years of above-target inflation, and despite scandalously low wages for early childhood educators, A 2021 report estimated a funding gap for the system of between $4.2 billion and $23.3 billion,[23] and demand for spaces is expected to run 220,000 ahead of supply in Ontario alone. For all the nice things I've said about how we can do industrial policy better, funding the child care gap and providing genuinely universal access seems to me to be a smarter investment in the economy than just about any business subsidy you can think of.

At the end of the day, any progressive government should see steady improvement of the quality of life of working people as its lodestar of legitimacy and political strength. Redistribution is part of that, but so is organizing for growth. Leaving Canada's coddled corporate class to steer our economy in partnership with a government unable to resist giving it virtually everything it wants is no longer tenable. To paraphrase David Lewis, the question is no longer whether we shall pursue industrial strategy or not. The question is whether we shall have it for monopoly by monopoly, or build a state capable of planning by the people for the people. A richer country and a more capable state able to take on the big challenges of our time are possible.

---

23 Andrea Mrozek, Peter Jon Mitchell, Brian Dijkema, *Look Before You Leap: The Real Costs and Complexities of National Daycare*, Cardus, May 6, 2021. https://www.cardus.ca/research/family/reports/look-before-you-leap/

And, honestly, between the post-COVID economic shocks, the new global fiscal and monetary policy environment, the climate crisis, and the ongoing challenge of reconciliation with Indigenous peoples, we're going to need a richer country and a more capable state. Canada will need every tool at its disposal to build an economy that continues to deliver good jobs, affordable homes, the healthcare and education systems we rely on, and dignified retirements for seniors.

NOTE ON SOURCES

In writing this piece, I made use of a lot of scholarly work as well as newspapers and primary sources. For Canadian economic and business history at a high level, I made heavy use of Michael Bliss' *Northern Enterprise* and Robin Naylor's History of *Canadian Business, 1876-1914*, and to a lesser extent Gerry Van Houten's *Corporate Canada: A Historical Outline*. Michael Atkinson and William Coleman's *The State, Business and Industrial Change in Canada* and Richard Harris' *Trade, Industrial Policy, and International Competition* were enormously helpful in providing historical data and analytical insight. Richard French and Richard Van Loon's *How Ottawa Decides* was very useful for the internal battles of the 1970s and 1980s. Shirley Tillotson's work informed the material on the development of taxation. Fred Lazar and John Lester's work were extremely helpful in granular looks at subsidy policies and counting up how much government actually spends. Don Nerbas' *Dominion of Capital* was very helpful for material on C.D. Howe. Gerhard Bassler's work on Joey Smallwood, Dimitriy Anastakis' on Richard Hatfield, and Janice MacKinnon's helpful memoir informed the sections on provincial misadventures. I also made heavy use of the MacDonald Commission report and the Nielsen task force reports. This is only a very partial inventory. I have tried to approach this project as rigorously as this short format allows – if you have questions on sourcing or data, please contact me.

ACKNOWLEDGEMENTS

This book is dedicated to my teachers, from high school all the way through to graduate school, who managed to take a curious but lazy student and help me find the things I'm passionate about. So a big thank you to Mr. Rosenzweig, Ms. Egashira, Dr. Torrance and Dr. Lundell, Dr. Whatmore, Dr. Wilson and Dr. Robson.

I would like to thank Ken Whyte at Sutherland House for asking me to write for Sutherland Quarterly and letting me run with this project. It was a surprise and a considerable honour to be asked—thank you for taking a chance on me.

I spoke with many people who informed my research and gave me advice about either the subject matter or about writing a book. Thank you to Nick Kadysh, Jim Balsillie, Fred Lazar, John Lester, Murad Hemmadi, David Moscrop, my old boss, Charlie Angus, and also to Nate Wallace, who read a first draft and offered me excellent, actionable feedback. Thank you, as well, to Etienne Rainville, my friend and podcasting partner, for your edits and for allowing me to occasionally miss games night to work.

More than anyone, I have to thank my wife, Miranda. Taking on this project while we were planning our wedding, and then picking it back up when you were pregnant, meant that you shouldered more than your fair share of that work while I spent evenings and weekends at the library or at my desk. You've been an incredible partner for over fourteen years, and I and Ambroise could not be luckier to have you in our lives.

ABOUT THE AUTHOR

Laurent Carbonneau is a policy professional working primarily on innovation, science and technology issues, with previous experience on Parliament Hill and national campaigns working for the NDP Leader's Office and a senior member of caucus. He has master's degrees from Carleton University (political management) and the University of St Andrews (history) and a BA from Mount Allison University. He lives in Ottawa with his family.

GIVE A <u>THOUGHTFUL</u> GIFT

1 YEAR PRINT & DIGITAL SUBSCRIPTION

Five Days of Hell in a Rocky Mountain Paradise
Jasper on Fire

The Rise and Rise of Canada's
Corporate Welfare Bums
At the Trough

Laurent Carbonneau

**SAVE 20% OFF THE $19.95
PER ISSUE COVER PRICE**

- **Four** print books
- **Free** home delivery
- Plus **four** eBooks
- **Free** digital access
to all SQ publications
- Automatic renewal

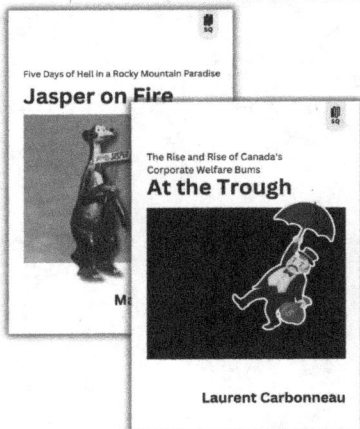

DELIVERY & PAYMENT DETAILS

Subscriber Info

NAME:
ADDRESS:
EMAIL: PHONE:

Payment Options

- Enclose a cheque or money order for $67.99 (includes HST) made out to
Sutherland House Inc. Send to Sutherland House, 304-416 Moore Ave,
Toronto, ON, Canada M4G 1C9
- Debit my Visa or MasterCard for $67.99 (includes HST)

CARD NUMBER: ____ ____ ____ ____   CVV: ___
EXPIRY DATE: __ / __   AMOUNT: $
PURCHASER'S NAME: SIGNATURE:

OR SUBSCRIBE ONLINE AT SUTHERLANDQUARTERLY.COM

GET THE <u>WHOLE</u> STORY

1 YEAR PRINT & DIGITAL SUBSCRIPTION

Five Days of Hell in a Rocky Mountain Paradise
Jasper on Fire

The Rise and Rise of Canada's
Corporate Welfare Bums
At the Trough

Laurent Carbonneau

SAVE 20% OFF THE $19.95 PER ISSUE COVER PRICE

- **Four** print books
- **Free** home delivery
- Plus **four** eBooks
- **Free** digital access
to all SQ publications
- Automatic renewal

DELIVERY & PAYMENT DETAILS

Subscriber Info

NAME:
ADDRESS:
EMAIL: PHONE:

Payment Options

- Enclose a cheque or money order for $67.99 (includes HST) made out to Sutherland House Inc. Send to Sutherland House, 304–416 Moore Ave, Toronto, ON, Canada M4G 1C9
- Debit my Visa or MasterCard for $67.99 (includes HST)

CARD NUMBER: ____ ____ ____ ____   CVV: ___
EXPIRY DATE: __ / __   AMOUNT: $
PURCHASER'S NAME: SIGNATURE:

OR SUBSCRIBE ONLINE AT SUTHERLANDQUARTERLY.COM

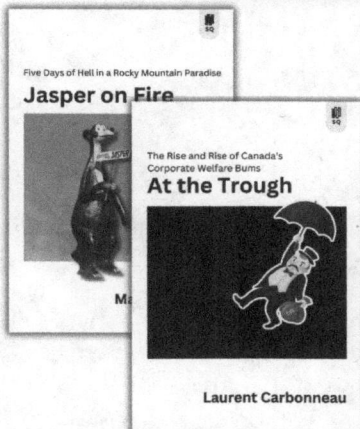